CONVEX SETS AND THEIR APPLICATIONS

Steven R. Lay

Department of Natural Sciences and Mathematics
Lee University
Cleveland, Tennessee

DOVER PUBLICATIONS, INC.
Mineola, New York

Copyright

Copyright © 1982 by Steven R. Lay
All rights reserved.

Bibliographical Note

This Dover edition, first published in 2007, is an unabridged republication of the work originally published in 1982 by John Wiley & Sons, Inc., New York.

Library of Congress Cataloging-in-Publication Data

Lay, Steven R., 1944–
    Convex sets and their applications / Steven R. Lay.
      p. cm.
    "This Dover edition, first published in 2007, is an unabridged republication of the work originally published in 1982 by John Wiley & Sons, Inc., New York."
    Includes bibliographical references and index.
    ISBN 0-486-45803-2 (pbk.)
      1. Convex sets. I. Title.

QA640.L38 2007
516'.08—dc22

2007000125

Manufactured in the United States of America
Dover Publications, Inc., 31 East 2nd Street, Mineola, N.Y. 11501

To
Frederick A. Valentine,
who first introduced me to
the beauty of a convex set

PREFACE

The study of convex sets is a fairly recent development in the history of mathematics. While a few important results date from the late 1800s, the first systematic study was the German book *Theorie der konvexen Körper* by Bonnesen and Fenchel in 1934. In the 1940s and 1950s many useful applications of convex sets were discovered, particularly in the field of optimization. The importance of these applications has in turn sparked a renewed interest in the theory of convex sets.

The relative youth of convexity has both good and bad implications for the interested student. Unlike most older branches of mathematics, there is not a vast body of background material that must be mastered before the student can reach significant unsolved problems. Indeed, even high school students can understand some of the basic results. Certainly upper level undergraduates have all the tools necessary to explore the properties of convex sets. This is good! Unfortunately, however, since the subject is so new, it has not yet "filtered down" beneath the graduate level in any comprehensive way. To be sure, there are several undergraduate books that devote two or three chapters to some aspects of convex sets, but there is no text at this level which has convex sets and their applications as its unifying theme. It is this void that the present book seeks to fill.

The mathematical prerequisites for our study are twofold: linear algebra and basic point-set topology. The linear algebra is important because we will be studying convex subsets of n-dimensional Euclidean space. The reader should be familiar with vectors and their inner product. The topological concepts we will encounter most frequently are open sets, closed sets, and compactness. Occasionally we will need to refer to continuity, convergence of sequences, or connectedness. These topics are typically covered in courses such as Advanced Calculus or Introductory Topology.

While the intent of this book is to introduce the reader to the broad scope of convexity, it certainly cannot hope to be exhaustive. The selection of topics has been influenced by the following goals: The material should be accessible to students with the aforementioned background; the material should lead the student to open questions and unsolved problems; and the material should highlight diverse applications. The degree to which the latter two goals are attained varies from chapter to chapter. For example, Chapter 4 on Kirchberger-type theorems includes few applications, but it does advance to the very edge of current mathematical research. On the other hand, the investigation of polytopes has progressed quite rapidly and so Chapter 8 can only be considered a brief introduction to the subject. Chapter 10 on optimization includes no unsolved problems, but has a major emphasis on applications.

For the past six years the material in this book has been used as a text in an upper-level undergraduate course in convexity. Since there is more material than can be covered in a three-unit semester course, the emphasis has varied from year to year depending on which topics were included. As a result of this flexibility, the material can be used as a geometry text for secondary teachers, as a resource for selected topics in applications, and as a bridge to higher mathematics for those continuing on to graduate school.

Throughout the book, the material is presented with the undergraduate student in mind. This is not intended to imply a lack of rigor, but rather to explain the inclusion of more examples and elementary exercises than would be typical in a graduate text. It is recognized that a course in convexity is more frequently found at the graduate level, but our attempt to make the material as clear and accessible as possible should not detract from its usefulness at this higher level as well. The exercises include not only computational problems, but also proofs requiring various levels of sophistication. The answers to many of the exercises are found at the back of the book, as are hints and references to the literature.

The first three chapters of the book form the foundation for all that follows. We begin in Chapter 1 by reviewing the fundamentals of linear algebra and topology. This enables us to establish our notation and leads naturally into the basic definition and properties of convex sets. Chapter 2 develops the important relationships between hyperplanes and convex sets, and this is applied in the following chapter to the theorems of Helly and Kirchberger.

Beginning with Chapter 4, the chapters become relatively independent of each other. Each chapter develops a particular aspect or application of convex sets. In Chapter 4 we look at several Kirchberger-type theorems in which the separating hyperplanes have been replaced by other geometrical figures. The results presented in this chapter are very recent, having only appeared in the literature in the last couple of years. Virtually any question that is posed beyond the scope of the text opens the door to an unsolved problem. (The answers to these will not be found in the back of the book!)

Chapter 5 investigates a number of topics which have had historical significance in the plane. After looking at sets of constant width and universal covers, we solve the classical isoperimetric problem under the assumption that a set of maximal area with fixed perimeter exists. In Chapter 6 we develop some of the tools necessary to prove the existence of such an extremal set.

There are many different ways to characterize convex sets in terms of local properties, and these are presented in Chapter 7. Chapter 8 introduces us to the basic properties of polytopes and gives special attention to how cubes,

PREFACE

pyramids, and bipyramids can be generalized to higher dimensions. In Chapter 9 we discuss two types of duality that occur in the study of convex sets: polarity and dual cones.

A third type of duality appears in Chapter 10, where we show how the theories of finite matrix games and linear programming are based on the separation properties of convex sets. The simplex method provides a powerful computational tool for solving a wide variety of problems. In Chapter 11 we apply the theory of convex sets to the study of convex functions.

A final word about the exercises is in order. They have been divided into three categories. (My students facetiously refer to them as the easy, the hard, and the impossible.) The first few exercises at the end of each section follow from the material in the text in a relatively straightforward manner. Some present examples illustrating the concepts, some ask for the details omitted in a proof, and others relate basic properties in a new way. The exercises which go considerably beyond the scope of the text are marked by an asterisk (*), although this distinction is, of course, sometimes arbitrary. The dagger (†) is used to denote exercises which are either open-ended in the questions they ask or are posing specific problems to which the answer is not currently known. Don't be afraid of the dagger! Only time will tell how difficult they may be. The solution of some of them undoubtedly will be found in a new insight or a moment of inspiration. They will lead you to the edge of current mathematical knowledge and challenge you to jump off on your own. I encourage you to accept this challenge. Good luck!

STEVEN R. LAY

Aurora, Illinois

ACKNOWLEDGMENTS

The writing and production of a book is a major undertaking, and it cannot be accomplished without the help of many people. Having a father and an older brother who are also mathematicians and authors is not always an easy burden to bear (at least for our wives), but it has proven to be a valuable resource, both for technical discussions and for encouragement. I have also appreciated the suggestions of several reviewers. In particular, the detailed comments of Professors Branko Grünbaum and Gerald Beer have been especially helpful. Thanks are due to my many former students for their ideas and criticisms as the project was developing, and to my wife, Ann, for her support throughout.

S.R.L.

CONTENTS

1. **Fundamentals**   1
   1. Linear Algebra and Topology, 1
   2. Convex Sets, 10

2. **Hyperplanes**   27
   3. Hyperplanes and Linear Functionals, 27
   4. Separating Hyperplanes, 33
   5. Supporting Hyperplanes, 41

3. **Helly-Type Theorems**   47
   6. Helly's Theorem, 47
   7. Kirchberger's Theorem, 55

4. **Kirchberger-type Theorems**   61
   8. Separation by a Spherical Surface, 61
   9. Separation by a Cylinder, 64
   10. Separation by a Parallelotope, 70

5. **Special Topics in E^2**   76
   11. Sets of Constant Width, 76
   12. Universal Covers, 84
   13. The Isoperimetric Problem, 88

6. **Families of Convex Sets**   94
   14. Parallel Bodies, 94
   15. The Blaschke Selection Theorem, 97
   16. The Existence of Extremal Sets, 101

7. **Characterizations of Convex Sets**   104
   17. Local Convexity, 104
   18. Local Support Properties, 107
   19. Nearest-Point Properties, 111

8. Polytopes — 116

20. The Faces of a Polytope, 116
21. Special Types of Polytopes and Euler's Formula, 123
22. Approximation by Polytopes, 133

9. Duality — 140

23. Polarity and Polytopes, 140
24. Dual Cones, 146

10. Optimization — 154

25. Finite Matrix Games, 154
26. Linear Programming, 168
27. The Simplex Method, 183

11. Convex Functions — 198

28. Basic Properties, 198
29. Support and Distance Functions, 205
30. Continuity and Differentiability, 214

Solutions, Hints, and References for Exercises — 222

Bibliography — 234

Index — 239

LIST OF NOTATION

R^n	n-dimensional linear space	1
E^n	n-dimensional Euclidean space	1
θ	the origin	2
$\varnothing$	the empty set	2
iff	if and only if	2
$\equiv$	equality (by definition)	2
R	the set of real numbers	2
$\langle x, y \rangle$	the inner product of x and y	2
$\|x\|$	the norm of x	2
$d(x, y)$	the distance from x to y	3
$B(x, \delta)$	the open ball of radius δ centered at x	4
$\sim S$	the complement of S	4
int S	the interior of S	5
cl S	the closure of S	5
bd S	the boundary of S	8
$\overline{xy}$	the line segment joining x and y	10
dim S	the dimension of S	12
relint S	the relative interior of S	12
relint $\overline{xy}$	the line segment $\overline{xy}$ without the endpoints	12
conv S	the convex hull of S	16
aff S	the affine hull of S	16
pos S	the positive hull of S	23
$[f : \alpha]$	$\{x \in \mathsf{E}^n : f(x) = \alpha\}$	27
$f(A) \leq \alpha$	$f(x) \leq \alpha$ for each $x \in A$	33
A_δ	the δ-parallel body to A	94
$D(A, B)$	the distance between A and B	95
$\mathcal{C}$	the set of all nonempty compact convex subsets of E^n	96

k-face	a k-dimensional face	116
$f_k(P)$	the number of k-faces of a polytope P	123
dc M	the dual cone of M	147
P_I and P_{II}	the first and second players in a matrix game	154
$E(x, y)$	the expected payoff for strategies x and y	157
$v(x)$	the value of a strategy x	158
v_I and v_{II}	the value of a game to P_I and P_{II}, respectively	158
epi f	the epigraph of f	199

1

FUNDAMENTALS

A study of convex sets can be undertaken in a variety of settings. The only necessary requirement is the presence of a linear structure, namely, a linear (or vector) space. Indeed, many interesting and useful results can be proved in this general context. At the other extreme are those concepts which seem to find their fulfillment in the Cartesian plane or three-space. We will chart a course somewhere in between these extremes and pursue our study of convex sets in n-dimensional Euclidean space. This setting is broad enough to include many of the important applications of convex sets, yet narrow enough to simplify many of the proofs. Very often the greatest difficulty in extending a result to n-dimensional spaces is encountered in going from two to three dimensions, and here our intuition is more reliable and we are aided by the ability to draw pictures. Whenever a new concept or result is presented, one should immediately construct examples in two and three dimensions to get a better "feeling" for the ideas involved.

SECTION 1. LINEAR ALGEBRA AND TOPOLOGY

The collection of all ordered n-tuples of real numbers (for $n = 1, 2, 3, \ldots$) can be made into the real linear space R^n by defining $(\alpha_1, \ldots, \alpha_n) + (\beta_1, \ldots, \beta_n) \equiv (\alpha_1 + \beta_1, \ldots, \alpha_n + \beta_n)$ and $\lambda(\alpha_1, \ldots, \alpha_n) \equiv (\lambda\alpha_1, \ldots, \lambda\alpha_n)$ for any n-tuples $(\alpha_1, \ldots, \alpha_n)$ and $(\beta_1, \ldots, \beta_n)$ and any real number λ. We define the inner product $\langle x, y \rangle$ of $x \equiv (\alpha_1, \ldots, \alpha_n)$ and $y \equiv (\beta_1, \ldots, \beta_n)$ to be the real number $\langle x, y \rangle \equiv \Sigma_{i=1}^n \alpha_i \beta_i$. The linear space R^n together with the inner product just defined is called n-dimensional Euclidean space, and is denoted by E^n. The n-tuples in E^n are referred to as points or vectors, interchangeably.

Throughout this book we will be dealing with subsets of n-dimensional Euclidean space unless otherwise indicated. If the particular dimension is important, it will be specified. Otherwise, the reader may assume that the context is E^n.

We will usually use lowercase letters such as x, y, z to denote points (or vectors) in E^n. Occasionally, x and y will be used as real variables when giving examples in E^2. For example we may write the linear equation $2x + 3y = 4$. The context should make the usage clear.

Capitals like A, B, C will denote subsets of E^n, and Greek letters such as α, β, δ will denote real scalars. The origin $(0, 0, \ldots, 0)$ will be denoted by θ, and the empty set by $\emptyset$. We will write $A \subset B$ to denote that A is a subset (proper or improper) of B. The shorthand "iff" means "if and only if," and the three-barred equal sign, $\equiv$, is used in defining a new point, set, or function. The set of real numbers will be denoted by R.

Our first theorem is a fundamental result from linear algebra, and its proof is left as an exercise.

1.1. Theorem. The inner product of two vectors in E^n has the following properties for all x, y, z in E^n:
  (a) $\langle x, x \rangle \geq 0$ and $\langle x, x \rangle = 0$ iff $x = \theta$.
  (b) $\langle x, y \rangle = \langle y, x \rangle$.
  (c) $\langle x + y, z \rangle = \langle x, z \rangle + \langle y, z \rangle$.
  (d) $\langle \alpha x, y \rangle = \alpha \langle x, y \rangle$ for every real α.

1.2. Definition. If $\langle x, y \rangle = 0$, then x and y are said to be **orthogonal** to each other.

By using the inner product in E^n we can talk about the "size" of a vector. Specifically, we define the norm of a vector as follows:

1.3. Definition. The **norm** of a vector x (denoted by $\|x\|$) is given by $\|x\| \equiv \langle x, x \rangle^{1/2}$. If $\|x\| = 1$, then x is called a **unit vector**.

The following properties of the norm are very useful:

1.4. Theorem. For all vectors x and y and real scalar α, the following hold:
  (a) $\|x\| > 0$ if $x \neq \theta$, and $\|\theta\| = 0$.
  (b) $\|\alpha x\| = |\alpha| \|x\|$.
  (c) $\|x + y\| \leq \|x\| + \|y\|$.
  (d) $\langle x, y \rangle = \|x\| \|y\| \cos \gamma$, where γ is the angle between the vectors x and y.

PROOF. Parts (a) and (b) follow directly from the definition, and part (c) follows from the Schwarz inequality:

$$\left(\sum_{i=1}^{n} \alpha_i \beta_i \right)^2 \leq \left(\sum_{i=1}^{n} \alpha_i^2 \right) \left(\sum_{i=1}^{n} \beta_i^2 \right).$$

Part (d) follows from the law of cosines. The details are left as an exercise. ∎

LINEAR ALGEBRA AND TOPOLOGY

By using the preceding norm we can define the distance between two points as follows:

1.5. Definition. If $x, y \in E^n$, then the **distance from** x **to** y, denoted by $d(x, y)$, is given by

$$d(x, y) \equiv \|x - y\|.$$

In terms of the inner product we have

$$d(x, y) = \langle x - y, x - y \rangle^{1/2},$$

and in terms of coordinates we have

$$d(x, y) = \left[\sum_{i=1}^{n} (\alpha_i - \beta_i)^2 \right]^{1/2},$$

where $x = (\alpha_1, \ldots, \alpha_n)$ and $y = (\beta_1, \ldots, \beta_n)$.

Examples

1. If $n = 1$, then the distance from x to y is just

$$\left[(x - y)^2\right]^{1/2} = |x - y|.$$

2. If $n = 2$, then the distance from x to y is given by the usual formula resulting from the Pythagorean Theorem. (See Figure 1.1.) The length of side a is $|\alpha_1 - \beta_1|$. The length of side b is $|\alpha_2 - \beta_2|$. Thus the length of the hypotenuse c is $d(x, y) \equiv \sqrt{(\alpha_1 - \beta_1)^2 + (\alpha_2 - \beta_2)^2}$.

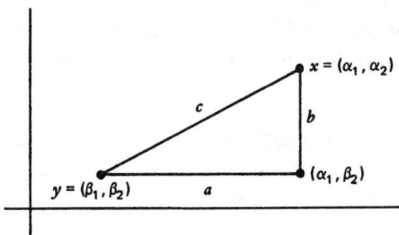

Figure 1.1.

1.6. Theorem. The distance function d has the following properties for all x, y, z in E^n:
(a) $d(x, y) \geq 0$ and $d(x, y) = 0$ iff $x = y$.
(b) $d(x, y) = d(y, x)$.
(c) $d(x, y) \leq d(x, z) + d(z, y)$.
(d) $d(\lambda x, \lambda y) = |\lambda| d(x, y)$ for every real λ.
(e) $d(x + z, y + z) = d(x, y)$.

PROOF. Parts (a) and (b) follow directly from Theorem 1.1. Part (c) is called the Triangle Inequality and says intuitively that the length of one side of a triangle is less than or equal to the sum of the lengths of the other two sides. (See Figure 1.2.) The proof of parts (c) and (d) follow from Theorem 1.4. Part (e) follows directly from the definition. ∎

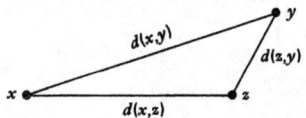

Figure 1.2.

Using the distance function we can define a topology for E^n just as we would for any other metric space.

1.7. Definition. For any $x \in \mathsf{E}^n$ and $\delta > 0$, the **open ball** $B(x, \delta)$ with center x and radius δ is given by

$$B(x, \delta) \equiv \{y \in \mathsf{E}^n : d(x, y) < \delta\}.$$

1.8. Definition. A point x is an **interior point** of the set S if there exists a $\delta > 0$ such that $B(x, \delta) \subset S$.

1.9. Definition. A set S is **open** if each of its points is an interior point of S.

1.10. Definition. The collection of all open subsets of E^n as defined above is called the usual **topology** for E^n. If S is a nonempty subset of E^n, then the **relative topology** on S is the collection of sets U such that $U = S \cap V$, where V is open in E^n.

It is easy to see that open balls, the whole space E^n, and the empty set $\emptyset$ are open sets. (See Exercise 1.5.) The union of any collection of open sets is an open set; the intersection of any finite collection of open sets is an open set.

1.11. Definition. A set S is **closed** if its complement $\sim S \equiv \mathsf{E}^n \sim S = \{x : x \in \mathsf{E}^n \text{ and } x \notin S\}$ is open.

LINEAR ALGEBRA AND TOPOLOGY

It is easy to see that all finite sets of points in E^n, the whole space E^n, and the empty set $\emptyset$ are closed sets. (See Exercise 1.6.) The intersection of any collection of closed sets is a closed set; the union of any finite collection of closed sets is a closed set. It is possible for a set to be neither open nor closed.

1.12. Definition. A set S is **bounded** if there exists a $\delta > 0$ such that $S \subset B(\theta, \delta)$.

Examples. Consider the following subsets of E^2:

$A = \{(x, y): (x - 4)^2 + (y + 2)^2 < 2\}$
$B = \{(x, y): 1 \leq x < 2 \text{ and } 1 < y \leq 3\}$
$C = \{(x, y): x + 2y \leq 4\}$

(See Figure 1.3.) The sets A and B are both bounded. Set C is not bounded. Set A is the open ball with center $(4, -2)$ and radius $\sqrt{2}$, and is an open set. The set C is closed, and set B is neither open nor closed.

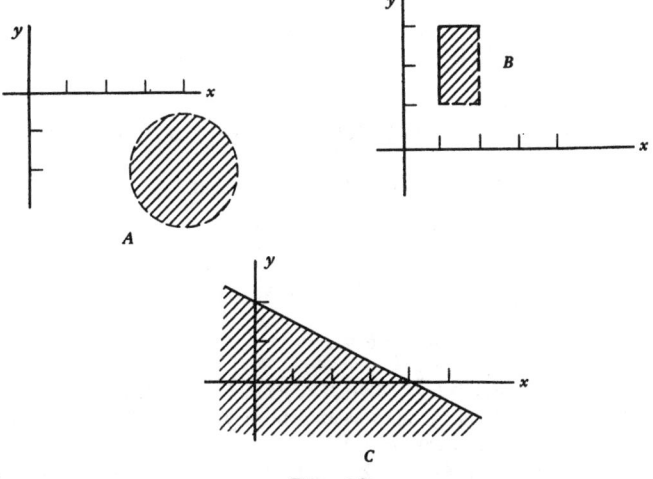

Figure 1.3.

1.13. Definition. The **interior** of a set S is the union of all the open sets contained in S. The **closure** of S is the intersection of all the closed sets containing S. The interior of S is denoted by int S and the closure of S by cl S.

It follows easily from the definitions that the interior of S is the set of all interior points of S. Also, a point x is in cl S iff for every $\delta > 0$, the open ball $B(x, \delta)$ contains at least one point of S.

1.14. Definition. A function $f: \mathsf{E}^n \to \mathsf{E}^m$ is **continuous** on E^n iff $f^{-1}(U)$ is an open subset of E^n whenever U is an open subset of E^m.

This definition of continuity is equivalent to the familiar ε-δ definition, which in terms of open balls becomes: f is continuous at $x \in \mathsf{E}^n$ iff for every $\varepsilon > 0$ there exists a $\delta > 0$ such that $f(B(x, \delta)) \subset B(f(x), \varepsilon)$. If f is continuous at each point of a set A, then f is continuous on A. We also recall that if $\{x_k\}$ is a sequence of points in E^n which converges to x and if f is continuous on E^n, then $\{f(x_k)\}$ converges to $f(x)$.

The fundamental theorem relating the linear and topological structures is the following:

1.15. Theorem. Each of the following functions is continuous:
  (a) $f: \mathsf{E}^n \times \mathsf{E}^n \to \mathsf{E}^n$ defined by $f(x, y) = x + y$.
  (b) For any fixed $a \in \mathsf{E}^n$, $f_a: \mathsf{E}^n \to \mathsf{E}^n$ defined by $f_a(x) = a + x$.
  (c) For any fixed $\lambda \in \mathsf{R}$, $f_\lambda: \mathsf{E}^n \to \mathsf{E}^n$ defined by $f_\lambda(x) = \lambda x$.
  (d) For any fixed $x, y \in \mathsf{E}^n$, $f: \mathsf{R} \to \mathsf{E}^n$ defined by $f(\lambda) = \lambda x + (1 - \lambda)y$.

PROOF. (a) Given $\varepsilon > 0$, let $\delta = \varepsilon/2$. Let (x_0, y_0) be a point in $\mathsf{E}^n \times \mathsf{E}^n$. Then for any x and y in E^n, if $d((x, y), (x_0, y_0)) < \delta$ we have

$$d((x, y), (x_0, y_0)) = \left\{[d(x, x_0)]^2 + [d(y, y_0)]^2\right\}^{1/2} < \delta$$

so that $d(x, x_0) < \delta$ and $d(y, y_0) < \delta$. It follows that

$$d(x + y, x_0 + y_0) \leq d(x + y, x + y_0) + d(x + y_0, x_0 + y_0)$$
$$= d(y, y_0) + d(x, x_0)$$
$$< \delta + \delta = \varepsilon.$$

Thus if $(x, y) \in B((x_0, y_0), \delta)$, then $f(x, y) \in B(f(x_0, y_0), \varepsilon)$ and f is continuous at (x_0, y_0). Since (x_0, y_0) was an arbitrary point in E^n, f is continuous.
  (b) This is just a special case of (a).
  (c) Let $\varepsilon > 0$ and let $x \in \mathsf{E}^n$. If $\lambda \neq 0$, let $\delta = \varepsilon/|\lambda|$. Then for any $y \in \mathsf{E}^n$ such that $d(x, y) < \delta$ we have

$$d(f_\lambda(x), f_\lambda(y)) = d(\lambda x, \lambda y) = |\lambda|\, d(x, y) < |\lambda|\, \frac{\varepsilon}{|\lambda|} = \varepsilon.$$

If $\lambda = 0$, then $d(f_\lambda(x), f_\lambda(y)) = d(\theta, \theta) = 0 < \varepsilon$ for any δ. In both cases we have $f_\lambda(B(x, \delta)) \subset B(f_\lambda(x), \varepsilon)$, and f_λ is continuous.
  (d) This proof is straightforward and is left as an exercise. ∎

LINEAR ALGEBRA AND TOPOLOGY

1.16. Definition. If $A, B \subset \mathsf{E}^n$ and $\lambda \in \mathsf{R}$, we define

$$A + B \equiv \{x + y : x \in A \text{ and } y \in B\}$$

$$\lambda B \equiv \{\lambda x : x \in B\}.$$

If A consists of a single point, $A \equiv \{x\}$, then we often write $x + B$ for $A + B$. The set $x + B$ is called a **translate** of B. The set λB is called a **scalar multiple** of B. If $\lambda \neq 0$, the set $x + \lambda B$ is said to be **homothetic** to B.

1.17. Theorem. Each set homothetic to an open set is open.

PROOF. For any $x \in \mathsf{E}^n$ and $\lambda \neq 0$, the function f given by $f(y) = x + \lambda y$ is continuous by Theorem 1.15. For $\lambda \neq 0$, its inverse $f^{-1}(z) = -(1/\lambda)x + (1/\lambda)z$ is also continuous. This implies that the original function maps open sets onto open sets. ∎

1.18. Corollary. Each set homothetic to a closed set is closed.

PROOF. For $\lambda \neq 0$, the function $f(y) = x + \lambda y$ is one-to-one. Thus $f(\sim A) = \sim f(A)$ for each $A \subset \mathsf{E}^n$ and the corollary follows immediately from the definition of a closed set. ∎

In trying to visualize the sum $A + B$ of two sets, it is often helpful to express the sum as a union of translates:

$$A + B = \bigcup_{x \in A} (x + B) = \bigcup_{y \in B} (A + y)$$

Examples
(See Figure 1.4.)

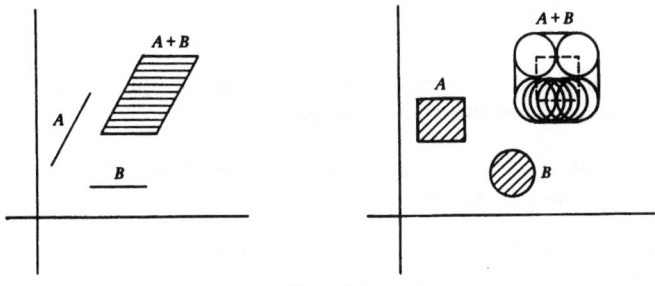

Figure 1.4.

1.19. Definition. The **boundary** of a set A, denoted by bd A, is defined by bd $A \equiv \text{cl } A \cap \text{cl}(\sim A)$.

In terms of open balls, a point x is in the boundary of A iff for every $\delta > 0$ the open ball $B(x, \delta)$ intersects both A and $\sim A$.

1.20. Definition. A subset A of E^n is said to be **compact** if it is closed and bounded.

There is an equivalent formulation of compactness in E^n which is based on the concept of an open cover: a collection of sets $\mathscr{F} = \{F_\alpha : \alpha \in \mathcal{C}\}$ is an **open cover** of S if each member of $\mathscr{F}$ is an open set and if S is contained in the union of all the members of $\mathscr{F}$.

1.21. Theorem (Heine-Borel). A subset S of E^n is compact iff for each open cover $\mathscr{F}$ of S there exist finitely many sets in $\mathscr{F}$ which by themselves form an open cover of S.

PROOF. See any standard analysis book such as Rudin (1976, p. 40). ∎

Compact sets enjoy a number of special properties not shared by other sets. In particular, compactness is preserved by continuous mappings. This in turn implies that continuous real-valued functions are bounded on compact sets and attain their maximum and minimum values. This is made precise in the following two theorems, the proofs of which are left as exercises.

1.22. Theorem. Let $f: \mathsf{E}^n \to \mathsf{E}^m$ be a continuous function and suppose A is a compact subset of E^n. Then $f(A) \equiv \{f(x): x \in A\}$ is a compact subset of E^m.

1.23. Theorem. Let A be a compact subset of E^n and let f be a real-valued continuous function defined on E^n. Then there exists $m \in \mathsf{R}$ such that $|f(x)| \leq m$ for all x in A. Furthermore, there exist points x_1 and x_2 in A such that

$$f(x_1) = \inf_{x \in A} f(x) \quad \text{and} \quad f(x_2) = \sup_{x \in A} f(x).$$

We conclude this first section by introducing the concept of connectedness. It will be of particular use to us in Section 17.

1.24. Definition. Two sets A and B are **separated** if both $A \cap \text{cl } B$ and $B \cap \text{cl } A$ are empty. A set S is **connected** if S is not the union of two nonempty separated sets. A subset C of S is a **component** of S if C is a maximal connected subset of S; that is, C is a connected subset of S such that if D is any connected subset of S that contains C, then $D = C$.

EXERCISES

It is easy to see that two separated sets must be disjoint, but disjoint sets need not be separated. For example, the intervals $A = [0, 2]$ and $B = (2, 4)$ in E^1 are not separated since $2 \in \text{cl } B$. However, the intervals $C = (0, 2)$ and B are separated. Correspondingly, $A \cup B$ is connected, while $C \cup B$ is not connected.

EXERCISES

1.1. Prove Theorem 1.1.

1.2. Finish the proof of Theorem 1.4.

1.3. Prove the following for all real α and all sets A, B, C:
  (a) $(A + B) + C = A + (B + C)$.
  (b) $\alpha(A + B) = \alpha A + \alpha B$.

1.4. In E^2, let A_1 be the closed line segment from the origin θ to the point $(2, 0)$. Let A_2 be the closed line segment from θ to $(0, 2)$. Let A_3 be the closed line segment from θ to $(2, 2)$. Let $A_4 = B(\theta, 1)$. Describe $A_1 + A_2$, $A_1 + A_4$, and $(A_1 + A_2) + A_3$.

1.5. Prove that each of the following is an open set.
  (a) An open ball.
  (b) E^n.
  (c) $\varnothing$.
  (d) The union of any collection of open sets.
  (e) The intersection of any finite collection of open sets.

1.6. Prove that each of the following is a closed set.
  (a) Any finite set.
  (b) E^n.
  (c) $\varnothing$.
  (d) The intersection of any collection of closed sets.
  (e) The union of any finite collection of closed sets.

1.7. Suppose $A \subset B$. Prove the following:
  (a) int $A \subset$ int B.
  ✓(b) cl $A \subset$ cl B.

1.8. Prove the following:
  (a) The interior of a set A is the set of all interior points of A.
  (b) A point x is in cl A iff for every $\delta > 0$, the open ball $B(x, \delta)$ contains at least one point of A.
  (c) cl $A = $ int $A \cup$ bd A and int $A \cap$ bd $A = \varnothing$.
  (d) A set A is open iff $A = $ int A.
  (e) A set A is closed iff $A = $ cl A.
  (f) cl $A = $ cl(cl A).
  (g) int $A = $ int(int A).

1.9. Prove that any finite set is compact.

1.10. For each statement below, either prove it is true or show it is false by a counterexample.
(a) If A is open, then for any set B, $A + B$ is open.
(b) If A and B are both closed, then $A + B$ is closed.

1.11. Verify the comments following Definition 1.14:
(a) A function $f: \mathsf{E}^n \to \mathsf{E}^m$ is continuous on E^n iff for each x in E^n and for each $\varepsilon > 0$ there exists a $\delta > 0$ such that $f(B(x, \delta)) \subset B(f(x), \varepsilon)$.
(b) If f is continuous and $\{x_k\}$ is a sequence which converges to x, then $\{f(x_k)\}$ converges to $f(x)$.

1.12. Prove Theorem 1.15(d).

1.13. Let p be a fixed point in E^n. Prove that the function $f: \mathsf{E}^n \to \mathsf{R}$ defined by $f(x) \equiv d(x, p)$ is continuous.

1.14. Let p be a fixed point in E^n with $p \neq \theta$. Prove that the function $f: \mathsf{E}^n \to \mathsf{R}$ defined by $f(x) \equiv \langle x, p \rangle$ is continuous.

1.15. Prove Theorem 1.22.

1.16. Prove Theorem 1.23.

1.17. For any nonempty set A and any point x we define the **distance from x to A** to be $d(x, A) \equiv \inf\{d(x, a): a \in A\}$. Prove that if A is closed and $x \notin A$, then $d(x, A) > 0$.

1.18. Let A and B be nonempty sets. We define the **distance from A to B** to be $d(A, B) \equiv \inf\{d(a, b): a \in A \text{ and } b \in B\}$.
(a) Give an example to show that it is possible for two disjoint closed sets A and B to satisfy $d(A, B) = 0$.
(b) Prove that if A is closed and B is compact and $A \cap B = \varnothing$, then $d(A, B) > 0$.

1.19. Let S be a connected set and suppose that K is a nonempty subset of S which is both open and closed relative to S. Prove that $K = S$.

1.20. Prove that if A is connected then cl A is connected.

1.21. Find an example of a set A such that cl A is connected, but A is not connected.

1.22. Let $f: \mathsf{E}^n \to \mathsf{E}^m$ be a continuous function and suppose that A is a connected subset of E^n. Prove that $f(A) \equiv \{f(x): x \in A\}$ is a connected subset of E^m.

SECTION 2. CONVEX SETS

Having surveyed the background of linear algebra and topology, we now direct our attention to the basic ideas of convex sets.

2.1. Definition. If x and y are points in E^n, the **line segment** $\overline{xy}$ joining x and y is the set of all points of the form $\alpha x + \beta y$ where $\alpha \geq 0$, $\beta \geq 0$ and $\alpha + \beta = 1$.

CONVEX SETS

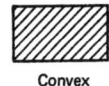  

Convex　　　　Star-shaped relative　　　Neither
　　　　　　　　　　to x

Figure 2.1.

2.2. Definition. A set S is **star-shaped relative to a point** $x \in S$ if for each point $y \in S$ it is true that $\overline{xy} \subset S$.

2.3. Definition. A set S is **convex** if for each pair of points x and y in S it is true that $\overline{xy} \subset S$.

It follows that a set is convex iff it is star-shaped relative to each point of the set. In fact, given any set S, the collection of those points with respect to which S is star-shaped always forms a convex subset of S. This is formalized in the following definition and theorem, with an intermediate lemma.

2.4. Definition. The **kernel** K of a set S is the set of all points $x \in S$ such that $\overline{xy} \subset S$ for each $y \in S$.

2.5. Lemma. Let x, y, z be three distinct points, and suppose $u \in \overline{xy}$, $u \neq x$, $u \neq y$. Then if $v \in \overline{zu}$, there exists a point $w \in \overline{zy}$ such that $v \in \overline{xw}$.

PROOF. As is often the case, a picture clarifies the problem, and in this case it makes the result intuitively obvious. (See Figure 2.2.)

Without loss of generality we may assume x is the origin θ. Since $u \in \overline{xy}$, $u \neq x$, $u \neq y$, $v \in \overline{zu}$, there exist real numbers λ ($0 < \lambda < 1$) and α ($0 \leq \alpha \leq 1$) such that $u = \lambda y$ and $v = \alpha z + (1 - \alpha)u$. It is easy to verify that

$$w \equiv \frac{\alpha}{\alpha + \lambda(1 - \alpha)} z + \frac{\lambda(1 - \alpha)}{\alpha + \lambda(1 - \alpha)} y \in \overline{zy}$$

and that $[\alpha + \lambda(1 - \alpha)]w = v$, so that $v \in \overline{xw}$. ∎

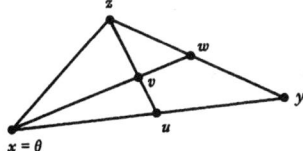

Figure 2.2.

2.6. Theorem. The kernel K of any set S is a convex set.

PROOF. Let $x \in K$, $y \in K$ with $x \neq y$, and let $u = \alpha x + (1 - \alpha)y$ for $0 < \alpha < 1$. We must show that $u \in K$. (See Figure 2.3.) To this end, let z be an

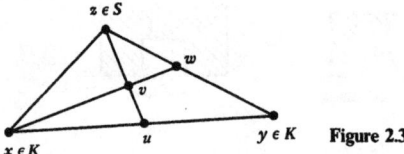
Figure 2.3.

arbitrary point in S. If $z = x$ or $z = y$, then $\overline{uz} \subset S$. If $z \neq x$ and $z \neq y$, let $v \in \overline{uz}$. By Lemma 2.5 there exists a point $w \in \overline{zy}$ such that $v \in \overline{xw}$. Since $y \in K$ and $z \in S$, $w \in S$. Since $x \in K$, $v \in S$. Hence for each $z \in S$ we have $\overline{uz} \subset S$, so $u \in K$. Therefore $\overline{xy} \subset K$ and K is convex. ∎

We recall from linear algebra that a finite set of points $x_1, x_2, \ldots, x_m$ in E^n is **linearly dependent** if there exist real numbers $\lambda_1, \lambda_2, \ldots, \lambda_m$, not all zero, such that $\lambda_1 x_1 + \lambda_2 x_2 + \cdots + \lambda_m x_m = \theta$. Otherwise, it is called **linearly independent**. (A sum of the form $\lambda_1 x_1 + \lambda_2 x_2 + \cdots + \lambda_m x_m$ is called a **linear combination** of the points $x_1, \ldots, x_m$.) A set of points B in E^n is a basis for a linear subspace L of E^n if B is a maximal linearly independent subset of L. Every linear subspace L has a basis, and all bases of L have the same number of elements. This common number of elements is called the dimension of L. The "natural" basis for E^n consists of the points $\varepsilon_1 \equiv (1, 0, 0, \ldots, 0)$, $\varepsilon_2 \equiv (0, 1, 0, \ldots, 0), \ldots, \varepsilon_n \equiv (0, 0, 0, \ldots, 1)$. Thus we find (as expected) that E^n has dimension n.

2.7. Definition. A translate of a linear subspace of E^n is called a **flat**. (Other names which are sometimes used include linear variety, variety, and affine subspace.) Two flats are **parallel** if one is a translate of the other. The **dimension of a flat** is the dimension of the corresponding parallel subspace. The **dimension of a set** S is the dimension of the smallest flat containing it, and is denoted by dim S. A flat of dimension 1 is called a **line**. A flat of dimension $n - 1$ is called a **hyperplane**.

In E^3 the proper linear subspaces consist of θ, the set of all lines through θ, and the set of all planes through θ. Thus the proper flats in E^3 are points (zero-dimensional), lines (one-dimensional) and planes (two-dimensional) which may or may not pass through the origin.

2.8. Notation. The interior of a set S relative to the minimal flat containing it is denoted by relint S.

If S consists of a single point x, then relint S = relint$\{x\} = x$. If S consists of a line segment $\overline{xy}$, then relint $\overline{xy}$ is the line segment without its endpoints. Specifically, relint $\overline{xy}$ consists of all points of the form $\alpha x + \beta y$ where $\alpha > 0$, $\beta > 0$, and $\alpha + \beta = 1$. We will prove later that every nonempty convex set has a nonempty relative interior. (See Corollary 2.28.)

CONVEX SETS

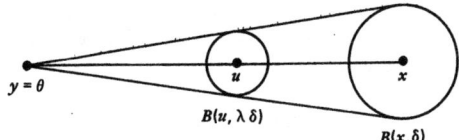

Figure 2.4.

2.9. Theorem. Let C be a convex set. If $x \in \text{int } C$ and $y \in C$, then $\text{relint } \overline{xy} \subset \text{int } C$.

PROOF. Choose $y \in C$ and without loss of generality assume $y = \theta$. (See Figure 2.4.) If $x \in \text{int } C$ then there exists an open ball $B(x, \delta) \subset C$. For any $u \in \text{relint } \overline{xy}$, there exists a λ ($0 < \lambda < 1$) such that $u = \lambda x$. Now $B(\lambda x, \lambda \delta) = \lambda B(x, \delta)$ (see Exercise 2.4), and since C is convex and $\theta \in C$, we have $\lambda B(x, \delta) \subset C$. Thus $B(u, \lambda \delta) \subset C$ and $u \in \text{int } C$. ∎

2.10. Corollary. If C is a convex set, then $\text{int } C$ is convex.

PROOF. Let $x, y \in \text{int } C$. Then by Theorem 2.9, $\text{relint } \overline{xy} \subset \text{int } C$. But $\overline{xy} = \text{relint } \overline{xy} \cup \{x, y\}$, so $\overline{xy} \subset \text{int } C$. ∎

2.11. Theorem. If C is a convex set, then $\text{cl } C$ is convex.

PROOF. Let x and y be in $\text{cl } C$ and let $u = \alpha x + \beta y$ where $\alpha \geq 0$, $\beta \geq 0$ and $\alpha + \beta = 1$. Furthermore, let $B(u, \delta)$ be an open ball centered at u. (See Figure 2.5.) Since x and y are in $\text{cl } C$, there exist points $x_0 \in B(x, \delta) \cap C$ and $y_0 \in B(y, \delta) \cap C$. We claim that $u_0 \equiv \alpha x_0 + \beta y_0 \in B(u, \delta)$. Indeed,

$$d(u, u_0) = d(\alpha x + \beta y, \alpha x_0 + \beta y_0)$$
$$\leq d(\alpha x + \beta y, \alpha x_0 + \beta y) + d(\alpha x_0 + \beta y, \alpha x_0 + \beta y_0)$$
$$= \alpha d(x, x_0) + \beta d(y, y_0)$$
$$< \alpha \delta + \beta \delta = (\alpha + \beta) \delta = \delta.$$

Since $u_0 \in \overline{x_0 y_0}$, we have $u_0 \in C$. Thus $u \in \text{cl } C$ and $\text{cl } C$ is convex. ∎

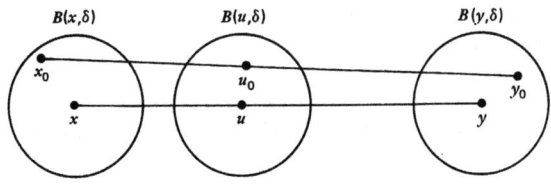

Figure 2.5.

In the definition of a convex set C we require that $\lambda x + (1 - \lambda)y \in C$ whenever $x, y \in C$ and $0 \leq \lambda \leq 1$. If we require that $\lambda x + (1 - \lambda)y \in C$ for all real λ, we have an "affine" set.

2.12. Definition. A set S is said to be **affine** if $x, y \in S$ implies that $\lambda x + (1 - \lambda)y \in S$ for all real numbers λ.

2.13. Theorem. A set S is affine iff it is a flat.

PROOF. Suppose that S is affine and let x be any fixed point in S. Let $U = -x + S$, so that $S = x + U$. We claim that U is a subspace of E^n. To prove this, let u_1 and u_2 be elements of U. Then there exist s_1 and s_2 in S such that $u_1 = -x + s_1$ and $u_2 = -x + s_2$. Thus for any real λ, we have

$$u_1 + \lambda u_2 = (-x + s_1) + \lambda(-x + s_2)$$
$$= -x + \lambda\left[2(\tfrac{1}{2}s_1 + \tfrac{1}{2}s_2) - x\right] + (1 - \lambda)s_1.$$

Since S is affine, $y \equiv \tfrac{1}{2}s_1 + \tfrac{1}{2}s_2$ is in S. But then $2y + (-1)x$ is in S and $\lambda[2y - x] + (1 - \lambda)s_1 \in S$, so $u_1 + \lambda u_2 \in -x + S = U$. Thus U is a subspace of E^n and $S = x + U$ is a flat.

Conversely, suppose $S = x + U$ for some $x \in \mathsf{E}^n$ and some subspace U. Let s_1 and s_2 be elements of S. Then there exist u_1 and u_2 in U such that $s_1 = x + u_1$ and $s_2 = x + u_2$. Thus for every real λ, we have

$$\lambda s_1 + (1 - \lambda)s_2 = \lambda(x + u_1) + (1 - \lambda)(x + u_2)$$
$$= x + \lambda u_1 + (1 - \lambda)u_2.$$

Since U is a subspace, $\lambda u_1 + (1 - \lambda)u_2 \in U$ and so $\lambda s_1 + (1 - \lambda)s_2 \in x + U = S$. Thus S is affine. ∎

2.14. Definition. Let $\lambda_i \in \mathsf{R}$ for $i = 1, 2, \ldots, k$, and suppose $\lambda_1 + \lambda_2 + \cdots + \lambda_k = 1$. Then

$$y \equiv \lambda_1 x_1 + \lambda_2 x_2 + \cdots + \lambda_k x_k$$

is called an **affine combination** of the points $x_1, x_2, \ldots, x_k$. If, in addition to the preceding conditions, we require that $\lambda_i \geq 0$ for all i, then y is called a **convex combination** of $x_1, x_2, \ldots, x_k$.

It follows readily that a set is convex (affine, respectively) iff it is closed under convex (affine, respectively) combinations of pairs of its elements. In fact, the next theorem shows that this is true for combinations of arbitrarily many elements.

CONVEX SETS

2.15. Theorem. A set S is convex iff every convex combination of points of S lies in S.

PROOF. The sufficiency of the condition is obvious from the definition. We will prove the necessity of the condition by an induction on the number m of points of S occurring in the convex combination. When $m = 2$, the conclusion follows from the definition. Assuming that every convex combination of k or fewer points of S yields a point of S, we consider a combination of $k + 1$ points: let $x = \lambda_1 x_1 + \cdots + \lambda_k x_k + \lambda_{k+1} x_{k+1}$ where $\lambda_1 + \cdots + \lambda_{k+1} = 1$, $\lambda_i \geq 0$ and $x_i \in S$ for all i. If $\lambda_{k+1} = 1$, then $x = x_{k+1}$, which belongs to S and there is nothing further to prove. Suppose that $\lambda_{k+1} < 1$. In this case $\lambda_1 + \cdots + \lambda_k = 1 - \lambda_{k+1} > 0$, and we have

$$x = (\lambda_1 + \cdots + \lambda_k) \times \left(\frac{\lambda_1}{\lambda_1 + \cdots + \lambda_k} x_1 + \cdots + \frac{\lambda_k}{\lambda_1 + \cdots + \lambda_k} x_k \right) + \lambda_{k+1} x_{k+1}.$$

By the induction hypothesis, the point

$$y \equiv \frac{\lambda_1}{\lambda_1 + \cdots + \lambda_k} x_1 + \cdots + \frac{\lambda_k}{\lambda_1 + \cdots + \lambda_k} x_k$$

belongs to S. Thus $x = (1 - \lambda_{k+1}) y + \lambda_{k+1} x_{k+1}$ is a convex combination of two points of S and so $x \in S$. ∎

2.16. Theorem. A set S is affine iff every affine combination of points of S lies in S.

PROOF. The proof is the same as in the previous theorem. ∎

2.17. Definition. A finite set of points $x_1, \ldots, x_m$ is **affinely dependent** if there exist real numbers $\lambda_1, \ldots, \lambda_m$, not all zero, such that $\lambda_1 + \cdots + \lambda_m = 0$ and $\lambda_1 x_1 + \cdots + \lambda_m x_m = \theta$. Otherwise it is **affinely independent**.

Note that in the preceding definition of affine dependence we have $\Sigma \lambda_i = 0$. This is consistent with Definition 2.14 for

$$x = \lambda_1 x_1 + \cdots + \lambda_k x_k \quad \text{with} \quad \sum_{i=1}^{k} \lambda_i = 1$$

iff

$$x - \lambda_1 x_1 - \cdots - \lambda_k x_k = \theta,$$

where this time the sum of all the coefficients on the left-hand side is zero.

Thus the concepts of affine and linear dependence are analogous, and this similarity is extended by the following theorem.

2.18. Theorem. Any subset of E^n consisting of at least $n + 1$ distinct points is linearly dependent. Any subset of E^n consisting of at least $n + 2$ distinct points is affinely dependent.

PROOF. Since E^n has dimension n, the maximal number of linearly independent points is n. Thus any subset of at least $n + 1$ distinct points is linearly dependent. To establish the second part of the theorem, suppose $x_1, \ldots, x_m$ are distinct points in E^n with $m \geq n + 2$. Then the $m - 1$ vectors $x_2 - x_1, x_3 - x_1, \ldots, x_m - x_1$ are linearly dependent. Thus there exist scalars $\alpha_2, \alpha_3, \ldots, \alpha_m$, not all zero, such that

$$\alpha_2(x_2 - x_1) + \alpha_3(x_3 - x_1) + \cdots + \alpha_m(x_m - x_1) = \theta.$$

That is, $\qquad -(\alpha_2 + \alpha_3 + \cdots + \alpha_m)x_1 + \alpha_2 x_2 + \cdots + \alpha_m x_m = \theta$

or $\qquad\qquad \alpha_1 x_1 + \alpha_2 x_2 + \cdots + \alpha_m x_m = \theta,$

where $\alpha_1 = -(\alpha_2 + \alpha_3 + \cdots + \alpha_m)$. Thus $\sum_{i=1}^{m}\alpha_i = 0$ and the points $x_1, x_2, \ldots, x_m$ are affinely dependent. ∎

We have now established three ways of taking combinations of points in E^n: linear, affine, and convex. If a set is closed* under all linear combinations, it is a **subspace**. If it is closed under affine combinations, it is a **flat**. If it closed under convex combinations, it is a **convex set**. We recall from linear algebra that the intersection of any collection of subspaces is itself a subspace. A similar result holds for affine sets and convex sets.

2.19. Theorem. If $\{S_\alpha\}$, $\alpha \in \mathcal{C}$, is any collection of convex sets, then $\bigcap_{\alpha \in \mathcal{C}} S_\alpha$ is convex. If $\{T_\beta\}$, $\beta \in \mathcal{B}$, is any collection of affine sets, then $\bigcap_{\beta \in \mathcal{B}} T_\beta$ is affine.

PROOF. If x and y are in $\bigcap S_\alpha$, then x and y are in each S_α. Since each S_α is convex, $\overline{xy} \subset S_\alpha$ for all α and hence $\overline{xy} \subset \bigcap S_\alpha$. The proof of the affine case is similar. ∎

2.20. Definition. The **convex hull** of a set S is the intersection of all the convex sets which contain S, and it is denoted by $\operatorname{conv} S$.

2.21. Definition. The **affine hull** of a set S is the intersection of all the affine sets which contain S, and it is denoted by $\operatorname{aff} S$.

---

*We are using closed here in the algebraic rather than the topological sense.

CONVEX SETS 17

Clearly S is convex iff $S = \text{conv}\, S$, and S is affine iff $S = \text{aff}\, S$. In a natural sense, $\text{conv}\, S$ is the "smallest" convex set containing S, and $\text{aff}\, S$ if the "smallest" affine set containing S. In view of the equivalence of affine sets and flats, we could have defined the dimension of S to be the dimension of $\text{aff}\, S$.

2.22. Theorem. For any set S, the convex (affine, respectively) hull of S consists precisely of all convex (affine, respectively) combinations of elements of S.

PROOF. We will prove the theorem for convex hulls; the proof for the affine case is similar. Let T denote the set of all convex combinations of elements of S. Since $\text{conv}\, S$ is convex and $S \subset \text{conv}\, S$, Theorem 2.15 implies that $T \subset \text{conv}\, S$. Conversely, let $x \equiv \alpha_1 x_1 + \cdots + \alpha_r x_r$ and $y \equiv \beta_1 y_1 + \cdots + \beta_s y_s$ be two elements of T. Then for any real λ with $0 \leq \lambda \leq 1$,

$$z \equiv \lambda x + (1-\lambda) y$$

$$= \lambda \alpha_1 x_1 + \cdots + \lambda \alpha_r x_r + (1-\lambda)\beta_1 y_1 + \cdots + (1-\lambda)\beta_s y_s$$

is an element of T since each coefficient is between 0 and 1 and

$$\sum_{i=1}^{r} \lambda \alpha_i + \sum_{j=1}^{s} (1-\lambda)\beta_j = \lambda \sum_{i=1}^{r} \alpha_i + (1-\lambda) \sum_{j=1}^{s} \beta_j = \lambda(1) + (1-\lambda)1 = 1.$$

Thus T is a convex set containing S and so $\text{conv}\, S \subset T$. Therefore, $\text{conv}\, S = T$. ∎

The following theorem is basic in the study of convex sets. It was first proved by Caratheodory in 1907. The previous theorem implies that a point x in the convex hull of S is a convex combination of (finitely many) points of S, but it places no restrictions on the number of points of S required to make the combination. Caratheodory's theorem says that in an n-dimensional space, the number of points of S in the convex combination never has to be more than $n + 1$.

2.23. Theorem (Caratheodory). If S is a nonempty subset of E^n, then every x in $\text{conv}\, S$ can be expressed as a convex combination of $n + 1$ or fewer points of S.

PROOF. Given a point x in $\text{conv}\, S$, Theorem 2.22 implies that $x = \lambda_1 x_1 + \cdots + \lambda_k x_k$ where $\lambda_1 + \cdots + \lambda_k = 1$, $\lambda_i \geq 0$ and $x_i \in S$ for all $i = 1, \ldots, k$. Our aim is to show that such an expression exists for x with $k \leq n + 1$.

If $k > n + 1$, then the points $x_1, \ldots, x_k$ are affinely dependent. Thus by Theorem 2.18 there exist scalars $\alpha_1, \ldots, \alpha_k$, not all zero, such that

$$\sum_{i=1}^{k} \alpha_i = 0 \quad \text{and} \quad \sum_{i=1}^{k} \alpha_i x_i = \theta.$$

Therefore, we have

$$\lambda_1 x_1 + \lambda_2 x_2 + \cdots + \lambda_k x_k = x$$

and

$$\alpha_1 x_1 + \alpha_2 x_2 + \cdots + \alpha_k x_k = \theta.$$

By subtracting an appropriate multiple of the second equation from the first, we will now eliminate one of the x_i and obtain a convex combination of fewer than k elements of S which is equal to x.

Since not all of the α_i are zero, we may assume (by reordering subscripts if necessary) that $\alpha_k > 0$ and that $\lambda_k/\alpha_k \leq \lambda_i/\alpha_i$ for all those i for which $\alpha_i > 0$. For $1 \leq i \leq k$, let $\beta_i = \lambda_i - (\lambda_k/\alpha_k)\alpha_i$. Then $\beta_k = 0$ and

$$\sum_{i=1}^{k} \beta_i = \sum_{i=1}^{k} \lambda_i - \frac{\lambda_k}{\alpha_k} \sum_{i=1}^{k} \alpha_i = 1 - 0 = 1.$$

Furthermore, each $\beta_i \geq 0$. Indeed, if $\alpha_i \leq 0$, then $\beta_i \geq \lambda_i \geq 0$. If $\alpha_i > 0$, then $\beta_i = \alpha_i(\lambda_i/\alpha_i - \lambda_k/\alpha_k) \geq 0$. Thus we have

$$\sum_{i=1}^{k-1} \beta_i x_i = \sum_{i=1}^{k} \beta_i x_i = \sum_{i=1}^{k} \left(\lambda_i - \frac{\lambda_k}{\alpha_k} \alpha_i \right) x_i$$

$$= \sum_{i=1}^{k} \lambda_i x_i - \frac{\lambda_k}{\alpha_k} \sum_{i=1}^{k} \alpha_i x_i = \sum_{i=1}^{k} \lambda_i x_i = x.$$

Hence we have expressed x as a convex combination of $k - 1$ of the points $x_1, \ldots, x_k$. This process may be repeated until we have expressed x as a convex combination of at most $n + 1$ of the points of S. ∎

Since the convex hull of a set S is the same as the set of all convex combinations of points of S, Caratheodory's Theorem can be rephrased as "x is in the the convex hull of S iff x is in the convex hull of $n + 1$ or fewer points of S." This result cannot, in general, be improved by decreasing the required number of points. Indeed, given any three noncollinear points in E^2, the centroid of the triangle formed by them is in the convex hull of all three, but is not in the convex hull of any two. It can be shown, however, (Exercise 2.35) that if S consists of at most n connected components in E^n then each point in

CONVEX SETS 19

conv S is a convex combination of n or fewer points of S. Further reductions in this direction do not seem to be possible.

We now introduce the general concept of a polytope, and in particular the simplest kind of polytope, the simplex. We will return to these topics in greater detail in Chapter 8, but our brief encounter here will be useful in proving that any nonempty convex set has a nonempty relative interior.

2.24. Definition. The convex hull of a finite set of points is called a **polytope** (or convex polytope). If $S = \{x_1, \ldots, x_{k+1}\}$ and $\dim S = k$, then conv S is called a *k*-**dimensional simplex**. The points $x_1, \ldots, x_{k+1}$ are called **vertices**.

Requiring that the dimension of $S = \{x_1, \ldots, x_{k+1}\}$ be equal to k is equivalent to requiring that the vectors $x_1, \ldots, x_{k+1}$ be affinely independent (i.e., the vectors $x_2 - x_1, \ldots, x_{k+1} - x_1$ must be linearly independent). (See Exercise 2.27.) A zero-dimensional simplex is a point; a one-dimensional simplex is a line segment; a two-dimensional simplex is a triangle; a three-dimensional simplex is a tetrahedron. From Theorem 2.22 we know that a point x in a simplex $S = \text{conv}\{x_1, \ldots, x_{k+1}\}$ can be expressed as a convex combination of the vertices. In fact, this representation is unique. Indeed, if

$$x = \sum_{i=1}^{k+1} \alpha_i x_i = \sum_{i=1}^{k+1} \beta_i x_i,$$

then $\sum_{i=1}^{k+1} \lambda_i x_i = \theta$, where $\lambda_i = \alpha_i - \beta_i$ $(1 \leq i \leq k+1)$. Since

$$\sum_{i=1}^{k+1} \lambda_i = \sum_{i=1}^{k+1} \alpha_i - \sum_{i=1}^{k+1} \beta_i = 1 - 1 = 0, \quad \lambda_1 = -(\lambda_2 + \cdots + \lambda_{k+1}).$$

Thus $\lambda_2(x_2 - x_1) + \cdots + \lambda_{k+1}(x_{k+1} - x_1) = \theta$; but $\dim S = k$, so $x_2 - x_1, \ldots, x_{k+1} - x_1$ are linearly independent. Therefore, $\lambda_i = 0$ for $2 \leq i \leq k+1$ (and hence also $\lambda_1 = 0$) and so $\alpha_i = \beta_i$ for all $i = 1, \ldots, k+1$. We have proved the following theorem:

2.25. Theorem. Let $S \equiv \{x_1, \ldots, x_{k+1}\}$ be a *k*-dimensional subset of E^n. Then each point in the simplex conv S has a unique representation as a convex combination of the vertices.

2.26. Definition. Suppose conv$\{x_1, \ldots, x_{k+1}\}$ is a *k*-dimensional simplex containing the point x. If the unique representation of x as a convex combination of the vertices is given by

$$x = \alpha_1 x_1 + \cdots + \alpha_{k+1} x_{k+1},$$

then the numbers $\alpha_1, \ldots, \alpha_{k+1}$ are called the **barycentric coordinates** of x. The point

$$x_0 = \frac{1}{k+1}(x_1 + \cdots + x_{k+1})$$

is called the **centroid** of the simplex.

By using the centroid we can prove that every nonempty convex set has a relative interior. We prove this first for simplices:

2.27. Theorem. Let $S \equiv \text{conv}\{x_1, \ldots, x_{k+1}\}$ be a k-dimensional simplex. Then relint $S \neq \emptyset$.

PROOF. Since the points $x_1, \ldots, x_{k+1}$ are affinely independent, each point in aff S has a unique representation as an affine combination of the x_i's. (See Exercise 2.28.) Thus we can define a function

$$f: \text{aff } S \to \mathsf{E}^{k+1} \quad \text{by} \quad f(\alpha_1 x_1 + \cdots + \alpha_{k+1} x_{k+1}) = (\alpha_1, \ldots, \alpha_{k+1}).$$

Let x_0 be the centroid of S. Then

$$f(x_0) = \left(\frac{1}{k+1}, \ldots, \frac{1}{k+1}\right).$$

Since the function f is continuous and each coordinate function is positive at x_0, it follows that there exists an open ball B around x_0 such that each coordinate function is positive for all x in $B \cap \text{aff } S$. But this implies that each x in $B \cap \text{aff } S$ is a convex combination of the points $x_1, \ldots, x_{k+1}$. Thus $(B \cap \text{aff } S) \subset \text{conv } S$ by Theorem 2.22, and x_0 is in the relative interior of S. ∎

2.28. Corollary. Let S be a k-dimensional convex subset of E^n. Then the relative interior of S is nonempty.

PROOF. Since S is k-dimensional, there exist $k+1$ affinely independent points $x_1, \ldots, x_{k+1}$ in S. Their convex hull is a k-dimensional simplex T, and by Theorem 2.27, relint $T \neq \emptyset$. Since relint $T \subset$ relint S, we have relint $S \neq \emptyset$. ∎

Sometimes it is helpful to use Caratheodory's Theorem as it relates to simplices: "x is in conv S iff x is in a simplex Δ whose vertices are points in S." Notice that this theorem says nothing about the interior of the sets. Even if we know that x is in the interior of conv S, we cannot conclude that x is in the interior of Δ. Indeed, if S consists of the four vertices of a square, then the

CONVEX SETS

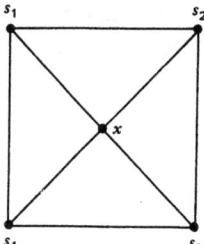

Figure 2.6.

center of the square is interior to $\text{conv}\,S$, but it is not interior to any triangle with vertices in S. (See Figure 2.6.) It does follow, however, that a point x in the convex hull of a set is always in the *relative* interior of a simplex of minimal dimension with vertices in the set.

This introductory chapter closes with a discussion of two theorems on the topological properties of the convex hull.

2.29. Theorem. If S is an open set, then $\text{conv}\,S$ is open.

PROOF. Since S is open, $S \cap \text{bd}\,\text{conv}\,S = \varnothing$. But $S \subset \text{conv}\,S$, so we must have $S \subset \text{int}\,\text{conv}\,S$. Furthermore, Corollary 2.10 implies that $\text{int}\,\text{conv}\,S$ is convex, so $\text{conv}\,S \subset \text{int}\,\text{conv}\,S$. On the other hand, $\text{int}\,\text{conv}\,S \subset \text{conv}\,S$, so $\text{conv}\,S = \text{int}\,\text{conv}\,S$, and $\text{conv}\,S$ is open. ∎

The convex hull of a closed set need not be closed. For example, in $\mathbf{E}^2$ let

$$S = \{(x, y): x^2 y^2 = 1 \text{ and } y > 0\}.$$

Then S is closed, but $\text{conv}\,S = \{(x, y): y > 0\}$ is not closed. If, however, we require S to be both closed and bounded, then the convex hull of S is both closed and bounded.

2.30. Theorem. If S is a compact set, then $\text{conv}\,S$ is compact.

PROOF. Let B be the compact subset of $\mathbf{E}^{n+1}$ defined by

$$B = \{(\alpha_1, \ldots, \alpha_{n+1}): \alpha_1 + \cdots + \alpha_{n+1} = 1 \text{ and } \alpha_i \geq 0 \text{ for } 1 \leq i \leq n+1\}.$$

The function f defined by

$$f(\alpha_1, \ldots, \alpha_{n+1}, x_{1,1}, \ldots, x_{1,n}, x_{2,1}, \ldots, x_{2,n}, \ldots, x_{n+1,1}, \ldots, x_{n+1,n})$$
$$= \sum_{i=1}^{n+1} \alpha_i (x_{i,1}, \ldots, x_{i,n})$$

is a continuous mapping of $E^{n+1} \times E^n \times \cdots \times E^n = E^{(n+1)^2}$ into E^n. By Caratheodory's Theorem 2.23, it maps the compact set $B \times S \times \cdots \times S$ onto conv S. Therefore, conv S is also compact. ∎

EXERCISES

2.1. Shade the kernel of each subset of E^2 shown in Figure 2.7.

2.2. Let A and B be convex sets.
  (a) Prove: $A \cap B$ is contained in the kernel of $A \cup B$.
  (b) Find an example to show that $A \cap B$ may not be equal to the kernel of $A \cup B$.

†2.3. Let K be an n-dimensional convex subset of E^n. Under what conditions will there exist a nonconvex set S such that K is the kernel of S?

2.4. Prove that $\lambda B(x, \delta) = B(\lambda x, \lambda \delta)$, where $\delta > 0$ and $\lambda > 0$.

2.5. If S is a convex set, prove that any set homothetic to S is convex.

2.6. If A and B are convex sets, prove that $A + B$ is convex.

2.7. Find the convex hull of the set of points (x, y) in E^2 which satisfy the given conditions:
  (a) $y = x^2$.
  (b) $y = x^2$ and $x \geq 0$.
  (c) $y = x^3$.
  (d) $y = 1/x$ and $x \geq \frac{1}{2}$.
  (e) $y = \sin x$.
  (f) $y = \tan x$ and $-\pi/2 < x < \pi/2$.

2.8. Prove that the open ball $B(x, \delta)$ is a convex set.

2.9. Prove that the convex hull of a bounded set is bounded.

2.10. Prove that a flat is a convex set.

2.11. Let S be a closed set with int $S \neq \emptyset$. If $x \notin S$ and $y \in$ int S, prove that there exists a point $z \in$ relint $\overline{xy}$ such that $z \in$ bd S.

2.12. Prove the following generalization of Theorem 2.9: Let S be a convex set. If $x \in$ int S and $y \in$ cl S, then relint $\overline{xy} \subset$ int S.

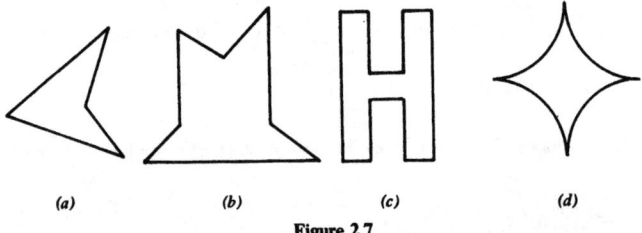

(a) (b) (c) (d)

Figure 2.7.

EXERCISES

2.13. Let S be a convex set with int $S \neq \emptyset$. Prove that cl(int S) = cl S, and find an example to show that the convexity of S is necessary.

2.14. Let S be a convex set with int $S \neq \emptyset$. Prove that int(cl S) = int S, and find an example to show that the convexity of S is necessary.

2.15. Let S be a convex set. Prove that bd(cl S) = bd S, and find an example to show that the convexity of S is necessary.

2.16. Let S be a convex set and let $\alpha > 0$, $\beta > 0$. Prove that $\alpha S + \beta S = (\alpha + \beta)S$, and find an example to show that the convexity of S is necessary.

2.17. Prove or give a counterexample: If A and B are nonempty convex sets with $A \subset B$, then relint $A \subset$ relint B.

2.18. If x and y are boundary points of the closed convex set S, prove that either $\overline{xy} \subset$ bd S or relint $\overline{xy} \subset$ int S.

2.19. Let S be a convex set with int $S \neq \emptyset$.
  (a) Prove or give a counterexample: The boundary of S is convex.
  (b) Prove or give a counterexample: The boundary of S is not convex.
  (c) Suppose that S is compact. Prove that the boundary of S is not convex.

2.20. Let S be a convex set with $x \in$ relint S and $y \neq x$. Prove that $y \in$ aff S iff relint $\overline{xy} \cap$ relint $S \neq \emptyset$.

2.21. Let A and B be nonempty sets. Prove the following:
  (a) If $A \subset B$, then $A \subset$ conv B.
  (b) If $A \subset B$ and B is convex, then conv $A \subset B$.
  (c) If $A \subset B$, then conv $A \subset$ conv B.
  (d) (conv A) $\cup$ (conv B) $\subset$ conv($A \cup B$).
  (e) conv($A \cap B$) $\subset$ conv $A \cap$ conv B.
  (f) conv(conv A) = conv A.
  (g) Find examples to show that equality need not hold in parts (d) and (e).

2.22. State and prove results analogous to Exercise 2.21 for the affine hull.

†2.23. Let S be a nonempty convex set and suppose A is a subset of S such that conv $A = S$. How are the sets A and S related if there are no other restrictions on S? Is there a unique "smallest" set whose convex hull is equal to S? What if S is open? What if S is closed? What if S is bounded? What if S is compact?

2.24. A point x is called a **positive combination** of the points $x_1, \ldots, x_m$ if $x = \alpha_1 x_1 + \cdots + \alpha_m x_m$, where all $\alpha_i \geq 0$. The set of all positive combinations of points of a set S is called the **positive hull** of S and is denoted by pos S. In E^2, let $S = \{(-1, 1), (1, 1)\}$. Find pos S.

2.25. Observe that in Exercise 2.24 we have pos $S \cap$ aff $S =$ conv S.
  (a) Show that this is not true in general by verifying the following example: Let $x_1 = (0, 1)$, $x_2 = (1, 1)$, $x_3 = (1, 0)$ and $x =$

(3, 2). Then $x \in \text{pos}\{x_1, x_2, x_3\} \cap \text{aff}\{x_1, x_2, x_3\}$ but $x \notin \text{conv}\{x_1, x_2, x_3\}$.
- (b) What special property does the set S in Exercise 2.24 have that makes $\text{pos } S \cap \text{aff } S = \text{conv } S$?

2.26. Let S be a nonempty set. Prove the following:
- (a) $\text{pos } S = \text{pos}(\text{conv } S)$.
- (b) If S is convex, then $x \in \text{pos } S$ iff $x = \lambda s$ for some $\lambda \geq 0$ and some $s \in S$.

2.27.
- (a) Prove that the set $\{x_1, \ldots, x_k\}$ is affinely dependent iff the set $\{x_2 - x_1, \ldots, x_k - x_1\}$ is linearly dependent.
- (b) Prove that the set $\{x_1, \ldots, x_k\}$ is affinely dependent iff one of the x_i's is an affine combination of the others.

2.28. √(a) Let $S \equiv \{x_1, \ldots, x_{k+1}\}$ be a k-dimensional set. Prove that each point p in aff S has a unique representation as an affine combination of $x_1, \ldots, x_{k+1}$. (The coefficients in this unique representation are called the **affine coordinates** of the point p with respect to S.)
- (b) Let $T \equiv \{(2, 0), (0, 5), (-1, 1)\}$ in E^2. Find the affine coordinates of $p_1 \equiv (2, 1)$, $p_2 \equiv (1, 1)$, $p_3 \equiv (1, \frac{1}{3})$, and $p_4 \equiv (1, 0)$ with respect to T.
- (c) Determine whether each of the points $p_1, \ldots, p_4$ in part (b) is interior to, on the boundary of, or outside of conv T.
- (d) Let $S \equiv \{x_1, x_2, x_3, x_4\}$ be a three-dimensional set. Consider the points whose affine coordinates with respect to S are given by $y_1 \equiv (2, 0, 0, -1)$, $y_2 \equiv (0, \frac{1}{2}, \frac{1}{4}, \frac{1}{4})$, $y_3 \equiv (\frac{1}{2}, 0, \frac{3}{2}, -1)$, and $y_4 \equiv (\frac{1}{3}, \frac{1}{4}, \frac{1}{4}, \frac{1}{6})$. Determine whether each of the points $y_1, \ldots, y_4$ is interior to, on the boundary of, or outside of conv S.

2.29. Suppose that F and G are k-dimensional flats with $F \subset G$. Prove that $F = G$.

2.30. In E^4, let $x_1 = (1, -1, 2, -1)$, $x_2 = (2, -1, 2, 0,)$, $x_3 = (1, 0, 2, 0)$ and $x_4 = (1, 0, 3, 1)$.
- (a) Show that the set $\{x_1, x_2, x_3, x_4\}$ is affinely independent.
- (b) Let $A \equiv \text{aff}\{x_1, \ldots, x_4\}$ and $B \equiv \{(\alpha_1, \ldots, \alpha_4): \alpha_1 + \alpha_2 + \alpha_3 - \alpha_4 = 3\}$. Show that $A = B$.

2.31.
- (a) Let F and G be flats. Prove that $F \cup G$ is convex iff $F \subset G$ or $G \subset F$.
- (b) Show by an example that the union of two arbitrary convex sets may be convex without either of the sets being a subset of the other.

2.32. Let $x_1, \ldots, x_k$ be mutually orthogonal nonzero vectors. Prove that they are linearly independent.

2.33. In E^2, let $x_1 = (1, 0)$, $x_2 = (1, 3)$, $x_3 = (4, 3)$ $x_4 = (4, 0)$, and $x = (\frac{7}{4}, \frac{5}{4})$. Then $x = \frac{1}{2}x_1 + \frac{1}{4}x_2 + \frac{1}{6}x_3 + \frac{1}{12}x_4$. Use the procedure in the proof of

Caratheodory's Theorem to express x as a convex combination of x_1, x_2, and x_3.

2.34. Prove the following variation of Caratheodory's Theorem: Let S be a nonempty subset of E^n, let $v \in S$, and let $x \in \text{conv}\,S$. Then there exist points $x_1, \ldots, x_r$ in S ($r \leq n$) such that $x \in \text{conv}\{v, x_1, \ldots, x_r\}$.

*2.35. Prove the following: Let S be a subset of E^n consisting of at most n connected components. If $x \in \text{conv}\,S$, then there exist $x_1, \ldots, x_k$ in S with $k \leq n$ such that $x \in \text{conv}\{x_1, \ldots, x_k\}$.

*2.36. Let S be a set consisting of m points ($m > n$) in E^n, and let $p \in \text{conv}\,S$. Caratheodory's Theorem 2.23 says that there exist $n + 1$ points in S which contain p in their convex hull, but it says nothing about how many different ways these points can be chosen. Prove the following: There are at least $\binom{m-n}{r-n}$ different selections of r of the points of S ($n < r \leq m$) that contain p in their convex hull.

2.37. Let A and B be nonempty convex sets. In this exercise we shall say that A encloses B (or equivalently that B is enclosed by A) if B is a proper subset of A and if $B \subset X \subset A$ where X is convex implies $X = B$ or $X = A$. That is, there are no convex sets "between" B and A.

(a) Find examples of convex sets that are enclosed by no convex set; exactly one convex set; exactly two convex sets; infinitely many convex sets.

(b) Find a sequence of convex sets $A_1, A_2, \ldots$ such that A_{i+1} encloses A_i for each $i = 1, 2, \ldots$.

(c) Prove or give a counterexample: If A and B are convex and A encloses B, then $A = B \cup \{p\}$ for some point $p \in \text{bd}\,B$.

(d) Prove or give a counterexample: If B is convex and $p \in (\text{bd}\,B) \sim B$, then $B \cup \{p\}$ is convex and encloses B.

*2.38. Find a convex subset S of a normed linear space E such that $\text{cl}\,S = E$ but $S \neq E$. (This is not possible when E is finite-dimensional.)

*2.39. Let S be a subset of E^n. Suppose that S contains at least three points that are not collinear, and suppose that for every three noncollinear points x, y, z in S there exists a unique point p in S such that $\overline{xp}$, $\overline{yp}$, and $\overline{zp}$ all lie in S. Prove that the kernel of S consists of a single point.

†2.40. Generalize Exercise 2.39 to find necessary and sufficient conditions that the set S be star-shaped with a k-dimensional kernel, $0 \leq k \leq n$.

*2.41. Let S be a nonempty compact subset of E^n. Prove that the following are equivalent:

(a) Every scalar multiple λS of S, with $0 < \lambda < 1$, can be expressed as the intersection of some family of translates of S.

(b) S is star-shaped and for every $x \in \mathsf{E}^n \sim S$, there exists a point y in the kernel of S such that $(\text{relint}\,\overline{xy}) \cap S = \varnothing$.

2.42. In applications of convexity to the theory of economics, it is sometimes useful to have a sufficient condition for the convex hull of a closed set to be closed.

*(a) Let S be a closed subset of the nonnegative orthant E^n_+ of E^n, and let e_i be the ith standard basis vector. Suppose that for any $x \in S$, there exist positive real numbers $\lambda_1, \ldots, \lambda_n$ such that $x + \lambda_i e_i \in S$, $i = 1, \ldots, n$. Prove that conv S is a closed set.

†(b) Find a sufficient condition for conv S to be closed whenever S is closed which does not require that $S \subset \mathsf{E}^n_+$.

*2.43. Let S be a closed nonconvex set. Prove that S is the union of two convex sets iff for each finite subset $x_1, \ldots, x_m$ of S, m odd, with the property that $\overline{x_i x_{i+1}} \not\subset S$ ($i = 1, \ldots, m - 1$), it follows that $\overline{x_1 x_m} \subset S$.

2
HYPERPLANES

Some of the most important applications of convex sets involve the problem of separating two convex sets by a hyperplane. In this chapter we will look carefully at the problem of separation and then use our results to obtain a characterization of closed convex sets.

SECTION 3. HYPERPLANES AND LINEAR FUNCTIONALS

In Chapter 1 we defined a hyperplane in E^n to be a flat of dimension $n - 1$. This is the natural generalization of the concept of a line in E^2 and a plane in E^3. Just as a line in E^2 can be described by an equation $\alpha x_1 + \beta x_2 = \delta$, and a plane in E^3 by an equation $\alpha x_1 + \beta x_2 + \gamma x_3 = \delta$, a hyperplane in E^n can be described by a similar equation.

3.1. Definition. A function $f: E^n \to E^m$ is said to be **linear** if it is additive and homogeneous. That is, for every $x, y \in E^n$ and $\lambda \in R$, we have

$$f(x + y) = f(x) + f(y)$$

and

$$f(\lambda x) = \lambda f(x).$$

A linear function from E^n into R is called a **linear functional**. For such a functional we denote the set $\{x \in E^n : f(x) = \alpha\}$ by $[f: \alpha]$.

3.2. Theorem. Suppose H is a subset of E^n. Then H is a hyperplane iff there exists a nonidentically zero linear functional f and a real constant δ such that $H = [f: \delta]$.

PROOF. Suppose first that H is a hyperplane and let $H_1 \equiv H - x_0$, where $x_0 \in H$. Then H_1 is an $n - 1$ dimensional subspace and so for any fixed point $y \in E^n \sim H_1$ we have $E^n = H_1 + R y$. Hence any $p \in E^n$ can be represented as $p = p_1 + \alpha y$, where $p_1 \in H_1$ and $\alpha \in R$. Furthermore, this representation is unique (see Exercise 3.2). Thus we can define a functional f by $f(p) = \alpha$. It follows that f is linear. Indeed, for $p = p_1 + \alpha y$ and $q = q_1 + \beta y$ we have

$$f(p + q) = f(p_1 + \alpha y + q_1 + \beta y)$$
$$= f[p_1 + q_1 + (\alpha + \beta)y] = \alpha + \beta = f(p) + f(q),$$

since $p_1 + q_1 \in H_1$ and $\alpha + \beta \in \mathsf{R}$. Furthermore, for any $\lambda \in \mathsf{R}$,

$$f(\lambda p) = f(\lambda p_1 + \lambda \alpha y) = \lambda \alpha = \lambda f(p)$$

since $\lambda p_1 \in H_1$ and $\lambda \alpha \in \mathsf{R}$. It follows that $[f: 0] = H_1$. But since f is linear and $H = H_1 + x_0$, we have $[f: \delta] = H$, where $\delta = f(x_0)$.

Conversely, suppose f is linear and that f is not identically zero on E^n. Then the dimension of the range of f is 1, so the dimension of the null space N of f must be $n - 1$. Given any $\alpha \in \mathsf{R}$, let x_0 be a point of E^n such that $f(x_0) = \alpha$. Then since f is linear, $[f: \alpha] = N + x_0$, so that $[f: \alpha]$ is a translate of an $n - 1$ dimensional subspace and thus is a hyperplane. ∎

Not only can a hyperplane be represented as the set of points where a linear functional assumes a specified value, but this representation is unique except for scalar multiples.

3.3. Theorem. If f and g are two linear functionals defined on E^n such that $[f: \alpha] = [g: \beta]$, then there exists a real constant $\lambda \neq 0$ such that $f = \lambda g$ and $\alpha = \lambda \beta$.

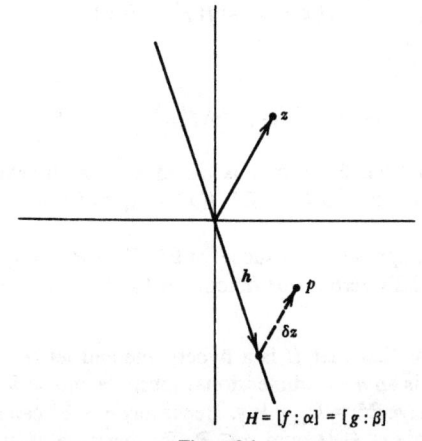

Figure 3.1.

HYPERPLANES AND LINEAR FUNCTIONALS

PROOF. Let $H = [f: \alpha] = [g: \beta]$. If either f or g is identically zero, then the conclusion holds trivially. Hence suppose $f \not\equiv 0$ and $g \not\equiv 0$. If $\alpha = 0$, then H is a hyperplane through θ and so $\beta = 0$. Similarly, if $\beta = 0$, then $\alpha = 0$. Thus suppose $\alpha = \beta = 0$, and choose $z \in \mathsf{E}^n \sim H$. Then $g(z) \neq 0$ and we can define $\lambda \equiv f(z)/g(z) \neq 0$. Let p be an arbitrary point in E^n. Since H is an $(n-1)$-dimensional subspace, we can write $p = h + \delta z$, where $h \in H$ and $\delta \in \mathsf{R}$. (See Figure 3.1.) Since $f(h) = g(h) = 0$, we have

$$f(p) - \lambda g(p) = [f(h) + \delta f(z)] - \lambda[g(h) + \delta g(z)]$$
$$= \delta[f(z) - \lambda g(z)] = 0.$$

Thus $f(p) = \lambda g(p)$ for all $p \in \mathsf{E}^n$.

Finally, suppose neither α nor β is zero, and define $\lambda \equiv \alpha/\beta$. Choose $z \in H$. Then since H is a hyperplane not containing θ, we have $\mathsf{E}^n = H + \mathsf{R}z$. Thus for any $p \in \mathsf{E}^n$, we can write $p = h + \delta z$, where $h \in H$ and $\delta \in \mathsf{R}$. (See Figure 3.2.) Then

$$f(p) - \lambda g(p) = [f(h) + \delta f(z)] - \lambda[g(h) + \delta g(z)]$$
$$= (1 + \delta)(\alpha - \lambda\beta) = 0,$$

since $f(h) = f(z) = \alpha$ and $g(h) = g(z) = \beta$. Hence $f(p) = \lambda g(p)$ for all $p \in \mathsf{E}^n$. ∎

One of the useful properties of linear functionals defined on E^n is that they are always continuous. (Linear functionals defined on an infinite dimensional

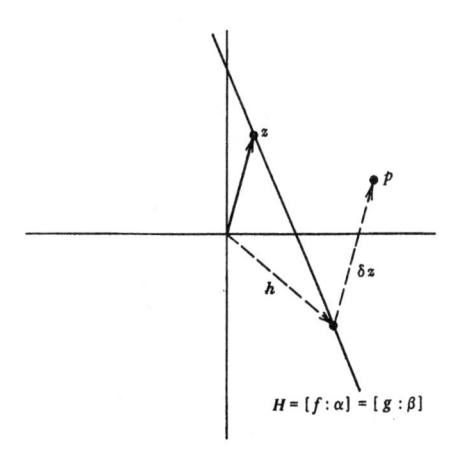

Figure 3.2.

topological linear space are not necessarily continuous.) This is proved in the following theorem, and it implies that hyperplanes (and hence all flats) are closed sets.

3.4. Theorem. Let $H \equiv [f: \gamma]$ be a hyperplane in E^n and let $\{e_1, \ldots, e_n\}$ be the standard basis for E^n. Then there exist real constants λ_i ($i = 1, \ldots, n$) such that for any $x \equiv \sum_{i=1}^n \alpha_i e_i$, $f(x) = \sum_{i=1}^n \alpha_i \lambda_i$. It follows that f is continuous and that H is closed.

PROOF. Let $\lambda_i \equiv f(e_i)$, $i = 1, \ldots, n$. Then

$$f(x) = f\left(\sum_{i=1}^n \alpha_i e_i\right) = \sum_{i=1}^n \alpha_i f(e_i) = \sum_{i=1}^n \alpha_i \lambda_i.$$

To see that f is continuous (in fact, uniformly continuous), let $\varepsilon > 0$ be given. Then for any two points

$$x \equiv \sum_{i=1}^n \alpha_i e_i \quad \text{and} \quad y \equiv \sum_{i=1}^n \beta_i e_i$$

in E^n for which

$$d(x, y) \leq \frac{\varepsilon}{\sum_{i=1}^n |\lambda_i|},$$

we have

$$|f(x) - f(y)| = \left|\sum_{i=1}^n \alpha_i \lambda_i - \sum_{i=1}^n \beta_i \lambda_i\right|$$

$$= \left|\sum_{i=1}^n \lambda_i (\alpha_i - \beta_i)\right|$$

$$\leq \sum_{i=1}^n |\lambda_i||\alpha_i - \beta_i|$$

$$\leq \frac{\varepsilon}{\sum_{i=1}^n |\lambda_i|} \sum_{i=1}^n |\lambda_i| = \varepsilon,$$

since $d(x, y) \geq |\alpha_i - \beta_i|$ for each $i = 1, \ldots, n$. Thus f is continuous, and this in turn implies that $f^{-1}(\{x: x > \gamma\})$ and $f^{-1}(\{x: x < \gamma\})$ are open subsets of E^n. Therefore, H, which is the complement of their union, is closed. ∎

HYPERPLANES AND LINEAR FUNCTIONALS

3.5. Corollary. If f is a linear functional defined on E^n, then there exists a point u in E^n such that $f(x) = \langle x, u \rangle$ for all x in E^n.

PROOF. Let $\{e_1, \ldots, e_n\}$ be the standard basis for E^n, and let $\lambda_i \equiv f(e_i)$, $i = 1, \ldots, n$. Then the point $u \equiv (\lambda_1, \ldots, \lambda_n) = \sum_{i=1}^{n} \lambda_i e_i$ has the property that for $x \equiv \sum_{i=1}^{n} \alpha_i e_i$ we have $f(x) = \sum_{i=1}^{n} \alpha_i \lambda_i = \langle x, u \rangle$. ∎

We have now established the correspondence between hyperplanes in E^n and linear functionals defined on E^n. Furthermore, Corollary 3.5 says that linear functionals can be represented by an inner product. Combining these two ideas we see that for any hyperplane H there exists a vector $u \neq \theta$ in E^n and a real number γ such that

$$H = \{x: \langle x, u \rangle = \gamma\}.$$

Conversely, given $u \neq \theta$ in E^n and a real number γ, the set described above is a hyperplane. If we let $x \equiv (x_1, \ldots, x_n)$ and $u \equiv (\lambda_1, \ldots, \lambda_n)$, then in the more familiar coordinate notation we have

$$H = \{(x_1, \ldots, x_n): \lambda_1 x_1 + \cdots + \lambda_n x_n = \gamma\}.$$

Since the vector u is orthogonal to any vector lying in the hyperplane H, (see Exercise 3.1), we make the following definition.

3.6. Definition. If the hyperplane H in E^n can be expressed as

$$H = \{x: \langle x, u \rangle = \gamma\},$$

then u is said to be a **normal** to H.

It follows readily that two hyperplanes are parallel iff their normals are scalar multiples of each other (Exercise 3.11).

3.7. Example. In E^2, let H be the hyperplane (line) corresponding to the equation $x + 2y = 4$. Then $H = [f: 4]$, where $f(x, y) = x + 2y$. Likewise, if $u \equiv (1, 2)$, then $H = \{(x, y): \langle (x, y), u \rangle = 4\}$. The subspace H_1 which is parallel to H is given by $[f: 0]$. If y is the vector $(1, -4)$, then the parallel hyperplane $H_2 = y + H$ is given by $[f: -3]$. (See Figure 3.3.)

In the preceding example, $[f: 4]$, $[f: 0]$, and $[f: -3]$ are all parallel. This is also true in general, as we see in the following theorem.

3.8. Theorem. Let $H_1 \equiv [f: \alpha]$ and $H_2 \equiv [f: \beta]$ be two hyperplanes in E^n corresponding to the same linear functional f. Then H_1 is parallel to H_2.

PROOF. Choose $h_1 \in H_1$ and $h_2 \in H_2$ and let $x_0 = h_2 - h_1$. We claim that

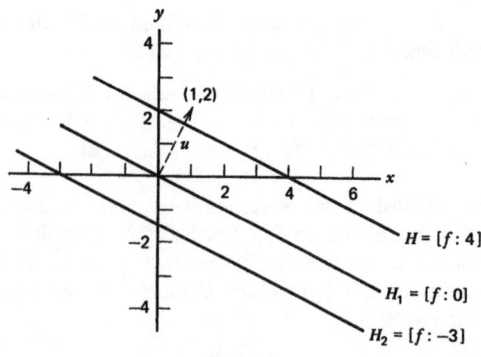

Figure 3.3.

$H_2 = x_0 + H_1$. For any element $x_0 + h$ in $x_0 + H_1$, we have

$$f(x_0 + h) = f(h_2 - h_1 + h) = \beta - \alpha + \alpha = \beta.$$

Thus $x_0 + h \in [f: \beta]$ and $x_0 + H_1 \subset H_2$.
Conversely, let $h \in H_2$ and write $h = x_0 + h - x_0 = x_0 + (h - h_2 + h_1)$. Since $f(h - h_2 + h_1) = \beta - \beta + \alpha = \alpha$, $h - h_2 + h_1 \in H_1$ and $h \in x_0 + H_1$. Thus $H_2 \subset x_0 + H_1$. ∎

EXERCISES

3.1. Let $H \equiv [f: \alpha]$ be a hyperplane in E^n and let u be a normal vector to H. Prove that u is orthogonal to any vector lying in H. That is, if x_1 and $x_2 \in H$, then u is orthogonal to $x_1 - x_2$.

3.2. Let V be an $(n-1)$-dimensional subspace of E^n and let $y \in \mathsf{E}^n \sim V$. Prove that each point $p \in \mathsf{E}^n$ has a unique representation as $p = x + \alpha y$, where $x \in V$ and $\alpha \in \mathsf{R}$.

3.3. Given parallel hyperplanes $H_1 = [f: \alpha]$ and $H_2 = x_0 + H_1$, where $x_0 \in \mathsf{E}^n$, prove that $H_2 = [f: \beta]$, where $\beta = f(x_0) + \alpha$.

3.4. In E^2, let H be the hyperplane $[f: 2]$, where $f(x, y) = 2x - 3y$.
  (a) Sketch the graphs of a normal vector u and the hyperplane H.
  (b) Let $(x_0, y_0) = (-1, 1)$. Sketch the graph of $(x_0, y_0) + H$ and find $\alpha \in \mathsf{R}$ such that $(x_0, y_0) + H = [f: \alpha]$.

3.5. Let V be a k-dimensional subspace ($0 \leq k \leq n-1$) of E^n and let $F_1 = x_1 + V$ and $F_2 = x_2 + V$ for points x_1, x_2 in E^n. Prove that either $F_1 = F_2$ or $F_1 \cap F_2 = \emptyset$. Thus two parallel flats either coincide or are disjoint.

SEPARATING HYPERPLANES

3.6. In E^2, let L be the line through the points $(-1, 2)$ and $(3, 1)$. Find a linear functional $f(x, y)$ and a real number α such that $L = [f: \alpha]$.

3.7. In E^4, let H be the hyperplane passing through the point $p \equiv (1, -3, 0, 2)$ having normal $u \equiv (2, 1, 4, -1)$. Which of the points $x_1 \equiv (0, 1, 1, 1)$, $x_2 \equiv (-2, 0, 1, 3)$, $x_3 \equiv (0, 1, -1, 2)$, and $x_4 \equiv (-1, 1, 1, 3)$ are on the same side of H as the origin, and which are not?

3.8. In E^4, let H be the hyperplane through the points $(1, 0, 1, 0)$, $(2, 3, 1, 0)$, $(1, 2, 2, 0)$, and $(1, 1, 1, 1)$.
   (a) Find a normal vector u.
   (b) Find a linear functional f and a real number α such that $H = [f: \alpha]$.

3.9. Prove: If F is a flat in E^n, then cl $F = F$. That is, flats are (topologically) closed sets.

3.10. Let S be a nonempty convex set. Prove the following:
   (a) cl $S \subset$ aff S.
   (b) aff(cl S) = aff S.
   (c) aff(relint S) = aff S.

3.11. Prove that two hyperplanes are parallel iff their normals are scalar multiples of each other.

3.12. Let f be a linear functional bounded above on a closed convex set S, and suppose that bd $S \neq \emptyset$. Show that

$$\sup_{x \in S} f(x) = \sup_{x \in \text{bd } S} f(x).$$

3.13. The three lines of a triangle divide E^2 into seven regions. State and prove a generalization of this for E^n.

3.14. Let $f: E^n \to E^m$ be a linear function. Prove the following:
   (a) If A is a convex subset of E^n, then $f(A) \equiv \{f(x): x \in A\}$ is a convex subset of E^m.
   (b) If B is a convex subset of E^m, then $f^{-1}(B) \equiv \{x: f(x) \in B\}$ is a convex subset of E^n.

SECTION 4. SEPARATING HYPERPLANES

Having characterized hyperplanes by linear functionals and given their representation by the inner product, we now consider the problem of "separating" two convex sets by a hyperplane. We begin with the necessary definitions and notation.

4.1. Notation. If f is a linear functional then $f(A) \leq \alpha$ means $f(x) \leq \alpha$ for each $x \in A$. Corresponding notations will be used when the inequalities are reversed or when they are strict.

4.2. Definition. The hyperplane $H \equiv [f: \alpha]$ **separates** (or **weakly separates**) two sets A and B if one of the following holds:
1. $f(A) \leq \alpha$ and $f(B) \geq \alpha$.
2. $f(A) \geq \alpha$ and $f(B) \leq \alpha$.

4.3. Definition. The hyperplane $H \equiv [f: \alpha]$ **strictly separates** two sets A and B if one of the following holds:
1. $f(A) < \alpha$ and $f(B) > \alpha$.
2. $f(A) > \alpha$ and $f(B) < \alpha$.

We notice at once that strict separation requires that the two sets be disjoint, while mere separation does not. Indeed, if two circles in the plane are externally tangent, then their common tangent line separates them (but does not separate them strictly). See Figure 4.1.

Although it is necessary that two sets be disjoint in order to strictly separate them, this condition is not sufficient, even for closed convex sets. Indeed, in E^2 let

$$A \equiv \left\{(x, y): x > 0 \text{ and } y \geq \frac{1}{x}\right\} \quad \text{and} \quad B \equiv \{(x, y): x \geq 0 \text{ and } y = 0\}.$$

Then A and B are disjoint closed convex sets, but they cannot be strictly separated by a hyperplane (line in E^2). Thus the problem of the existence of a hyperplane separating (or strictly separating) two sets is more complex than it might appear to be at first.

4.4. Lemma. Let S be an open convex subset of E^2. If $x \in E^2 \sim S$, then there exists a line L containing x that is disjoint from S.

PROOF. We may assume without loss of generality that x is the origin θ. Since $\theta \notin S$ and S is open, the positive hull of S (see Exercise 2.24) forms an open sector with vertex θ. Since S is convex, the angle δ of this sector is less than or equal to π. It follows that a line L containing one side of the sector will be disjoint from S. (See Figure 4.2.) ∎

Figure 4.1.

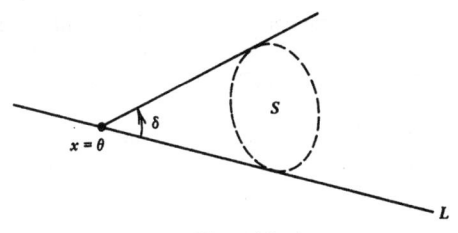

Figure 4.2.

4.5. Theorem. Let F be a k-dimensional flat and S an open convex set in E^n such that $F \cap S = \varnothing$. If $0 \leq k \leq n - 2$, then there exists a flat F^* of dimension $k + 1$ such that $F^* \supset F$ and $F^* \cap S = \varnothing$.

PROOF. Suppose $0 < k \leq n - 2$ and let $\{e_1, \ldots, e_n\}$ be the standard basis for E^n. We may assume without loss of generality that $\theta \in F$ and that F is spanned by $\{e_1, \ldots, e_k\}$. Let G be the $(n - k)$-dimensional subspace spanned by $\{e_{k+1}, \ldots, e_n\}$ and project E^n onto G by the mapping

$$\pi(\alpha_1, \ldots, \alpha_n) = (0, \ldots, 0, \alpha_{k+1}, \ldots, \alpha_n).$$

It follows (Exercise 4.3) that $\pi(S)$ is a relatively open convex subset of G and that $\theta \notin \pi(S)$. Let P be the two-dimensional subspace (plane through θ) spanned by $\{e_{n-1}, e_n\}$. Then $P \subset G$ and $P \cap \pi(S)$ is a relatively open convex subset of the plane P that does not contain the origin. By Lemma 4.4, there exists a one-dimensional subspace (line through θ) L contained in P such that $L \cap \pi(S) = \varnothing$. But then $F^* \equiv \pi^{-1}(L)$ is a $(k + 1)$-dimensional subspace of E^n such that $F^* \supset F$ and $F^* \cap S = \varnothing$. (See Figure 4.3.)

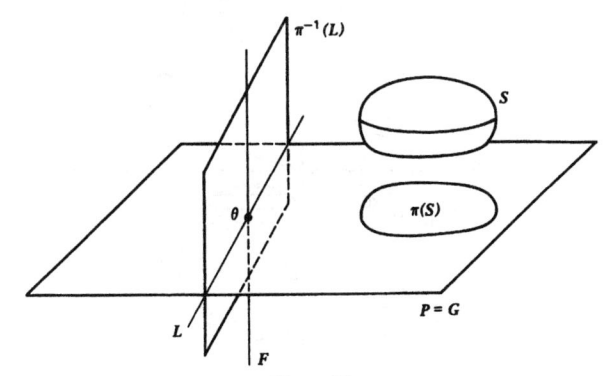

Figure 4.3.

If $k = 0$, let P be any plane containing the flat (point) F. Then $S \cap P$ is a relatively open convex subset of P and Lemma 4.4 implies the existence of a line F^* in P such that $F^* \supset F$ and $F^* \cap S = \varnothing$. ∎

4.6. Corollary: Let S be an open convex subset of E^n and let F be a k-dimensional flat ($0 \leq k < n$). If $F \cap S = \varnothing$, then there exists a hyperplane H such that $H \supset F$ and $H \cap S = \varnothing$.

PROOF. Apply Theorem 4.5 to F a total of $n - k - 1$ times. Each time the flat will be increased in dimension by one, until finally it is a hyperplane. ∎

4.7. Theorem. Suppose A and B are convex subsets of E^n. If int $A \neq \varnothing$ and $B \cap \text{int } A = \varnothing$, then there exists a hyperplane H which separates A and B.

PROOF. Suppose first that A is an open set. Then the set $A - B$ is open and convex. (See Exercises 1.10 and 2.6.) Since $A \cap B = \varnothing$, $\theta \notin A - B$, and by Corollary 4.6 there exists a hyperplane $H \equiv [f: 0]$ containing θ such that $H \cap (A - B) = \varnothing$. (See Figure 4.4.) Since $A - B$ is convex, we may assume without loss of generality that $f(A - B) > 0$. Then for any $x \in A$ and $y \in B$, we have

$$0 < f(x - y) = f(x) - f(y),$$

so $f(x) > f(y)$. It follows that $\{f(x): x \in A\}$ is bounded below so that $\alpha \equiv \inf\{f(x): x \in A\}$ is finite. Then A and B are separated by the hyperplane $[f: \alpha]$.

If A is not open, then we can apply the preceding argument to int A and obtain a hyperplane $H \equiv [f: \alpha]$ such that $f(\text{int } A) \geq \alpha$ and $f(B) \leq \alpha$. But $f(\text{int } A) \geq \alpha$ implies $f(A) \geq \alpha$, so H also separates A and B. ∎

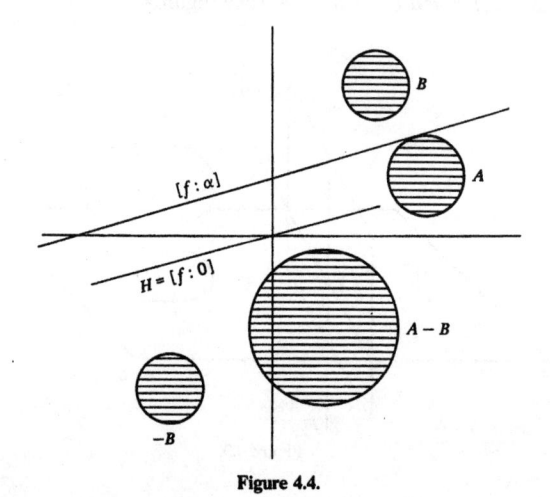

Figure 4.4.

SEPARATING HYPERPLANES

It is easy to see that the requirement in Theorem 4.7 that one of the sets must have an interior cannot be eliminated. Indeed, in $\mathbf{E}^3$ let $A = \{(x, y, z): |x| \leq 1, |y| \leq 1, \text{ and } z = 0\}$ and $B = \{(x, y, z): x = 0, y = 0, \text{ and } |z| \leq 1\}$. Then int $A = \emptyset$ so $B \cap$ int $A = \emptyset$, but no hyperplane can separate A and B. It is possible, however, to weaken the requirement. The next theorem gives the best possible characterization of being able to separate two convex sets. It is preceded by two basic definitions and a lemma.

4.8. Definition. Given a hyperplane $H \equiv [f: \alpha]$, the sets

$$\{x: f(x) \geq \alpha\} \quad \text{and} \quad \{x: f(x) \leq \alpha\}$$

are called the **closed half-spaces** determined by H. If the inequalities are replaced by strict inequalities, the resulting sets are called the **open half-spaces** determined by H.

4.9. Definition. The hyperplane $H \equiv [f: \alpha]$ is said to **bound** the set S if either $f(S) \geq \alpha$ or $f(S) \leq \alpha$ holds. If H does not bound the set S, then it is said to **cut** S. In other words, H cuts S if there exist points x, y in S such that $f(x) < \alpha$ and $f(y) > \alpha$.

4.10. Lemma. A hyperplane $H \equiv [f: \alpha]$ cuts a convex set S iff the following two conditions hold:
1. $H \not\supset S$.
2. $H \cap$ relint $S \neq \emptyset$.

PROOF. Suppose that H cuts S. Then there exist points x, y in S such that $f(x) < \alpha$ and $f(y) > \alpha$. Since $f(H) = \alpha$, (1) is true. Next let $z \in$ relint S. By Theorem 2.9 applied to S (considering S as a subset of aff S), all the points of $\overline{xz}$ and $\overline{yz}$, except possibly x and y, belong to relint S. Since $\overline{xz} \cup \overline{yz}$ is connected and f is continuous, there exists a point p in $\overline{xz} \cup \overline{yz}$ such that $f(p) = \alpha$. Since p is neither x nor y, it is in relint S, and (2) is proved.

Conversely, suppose (1) and (2) are satisfied. We may assume without loss of generality that $\theta \in H \cap$ relint S and that there exists a point x in $S \sim H$ such that $f(x) > 0$. Since $\theta \in$ relint S, there exists a $\delta > 0$ such that $-\delta x \in S$. But then $f(-\delta x) = -\delta f(x) < 0$ so H cuts S. ∎

4.11. Theorem. Suppose A and B are convex subsets of $\mathbf{E}^n$ such that $\dim(A \cup B) = n$. Then A and B can be separated by a hyperplane iff relint $A \cap$ relint $B = \emptyset$.

PROOF. Suppose there exists a point x in relint $A \cap$ relint B. If H is a hyperplane that separates A and B, then certainly H contains the point x. Since $\dim(A \cup B) = n$, it is not possible for both A and B to be contained in H. Thus Lemma 4.10 implies that H cuts one of the sets, a contradiction to their being separated by H. Therefore, no hyperplane can separate A and B.

To prove the converse we first note that a hyperplane will separate A and B iff it separates relint A and relint B. Thus we may assume that A and B are relatively open sets, so that $A \cap B = \emptyset$. The rest of the proof is by induction on the dimension of the set A, in descending order.

If $\dim A = n$, then int $A \neq \emptyset$, and Theorem 4.7 implies the existence of a separating hyperplane. Suppose the theorem is true whenever one of the sets has dimension greater than or equal to k ($1 \leq k \leq n$), and suppose $\dim A = k - 1$. To simplify notation we assume that $\theta \in A$. Let $H \equiv [f: 0]$ be a hyperplane containing A and let $x \in \mathbf{E}^n$ such that $f(x) > 0$. Define

$$C \equiv \text{conv}[A \cup (A + x)] \quad \text{and} \quad D \equiv \text{conv}[A \cup (A - x)].$$

(See Figure 4.5.) If $B \cap C \neq \emptyset$ and $B \cap D \neq \emptyset$, then the convexity of B and $C \cup D$ would imply $B \cap A \neq \emptyset$, a contradiction. Hence, without loss of generality, suppose $B \cap C = \emptyset$. Since $\dim C = k$, our induction assumption implies that there exists a hyperplane H separating C and B. Since $A \subset C$, H also separates A and B, and the induction is complete. ∎

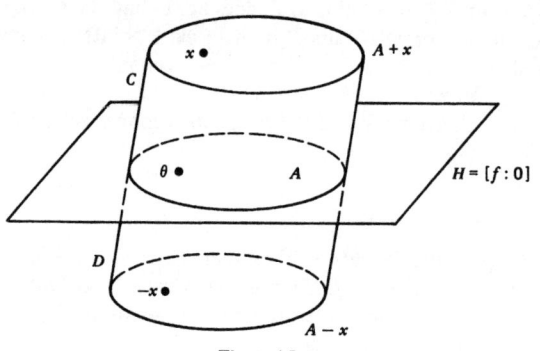

Figure 4.5.

It should be noted that requiring $\dim(A \cup B) = n$ is essential in Theorem 4.11. If $\dim(A \cup B) < n$, then $A \cup B$ is contained in a hyperplane, and this hyperplane separates A and B regardless of whether relint A and relint B are disjoint. Such a separation is of little value, however, and is said to be improper.

We now turn to the problem of strictly separating two convex sets by a hyperplane. It is easy to construct examples of two disjoint noncompact sets that cannot be strictly separated. If the sets are both compact, however, then such a separation is always possible. In fact, the following slightly stronger theorem holds:

4.12. Theorem. Suppose A and B are nonempty convex sets such that A is compact and B is closed. Then there exists a hyperplane H which strictly separates A and B iff A and B are disjoint.

SEPARATING HYPERPLANES 39

PROOF. Suppose first that A and B are disjoint and let $d(A, B) \equiv \inf\{d(x, y): x \in A \text{ and } y \in B\}$. It follows (see Exercise 1.18) that $\delta = d(A, B) > 0$. Let $S \equiv B(\theta, \delta/2)$ be the open ball of radius $\delta/2$ centered at the origin. Then $A + S$ and $B + S$ are disjoint open convex subsets of E^n, so Theorem 4.11 implies that there exists a hyperplane H which separates $A + S$ and $B + S$. Clearly H strictly separates A and B. The converse is immediate. ∎

Several extensions and reformulations of Theorem 4.12 are of particular interest. For example, by dropping the requirement that A and B be convex, we obtain the following:

4.13. Theorem. Suppose A and B are nonempty compact sets. Then there exists a hyperplane H which strictly separates A and B iff $\text{conv}\, A \cap \text{conv}\, B = \varnothing$.

PROOF. Suppose that $\text{conv}\, A \cap \text{conv}\, B = \varnothing$. Since the convex hull of a compact set is compact, we may apply Theorem 4.12 to obtain a hyperplane H that strictly separates $\text{conv}\, A$ and $\text{conv}\, B$. Clearly H also strictly separates the smaller sets A and B.

Conversely, suppose the hyperplane $H \equiv [f: \alpha]$ strictly separates A and B. We may assume without loss of generality that $f(A) < \alpha$ and $f(B) > \alpha$. Let $x \equiv \lambda_1 x_1 + \cdots + \lambda_k x_k$ be any convex combination of elements of A. Then

$$f(x) = \lambda_1 f(x_1) + \cdots + \lambda_k f(x_k) < \lambda_1 \alpha + \cdots + \lambda_k \alpha = \alpha,$$

since $\lambda_1 + \cdots + \lambda_k = 1$. Thus $f(\text{conv}\, A) < \alpha$. Likewise $f(\text{conv}\, B) > \alpha$, so $H \equiv [f: \alpha]$ strictly separates $\text{conv}\, A$ and $\text{conv}\, B$. By Theorem 4.12, $\text{conv}\, A$ and $\text{conv}\, B$ must be disjoint. ∎

As a special case of Theorem 4.13 we have the following: "A point x can be strictly separated from a compact set S by a hyperplane iff $x \notin \text{conv}\, S$." In light of Theorem 2.23 (Caratheodory) this can be restated as: "A point x can be strictly separated from a compact set S in E^n by a hyperplane iff for each subset T of $n + 1$ or fewer points of S, there exists a hyperplane strictly separating x and T." The application of this result to two compact sets is formalized in the following theorem. The details of the proof are left to the reader.

4.14. Theorem. Suppose A and B are nonempty compact subsets of E^n. Then there exists a hyperplane strictly separating A and B iff for each subset T of $n + 1$ or fewer points of B there exists a hyperplane strictly separating A and T.

PROOF. Exercise 4.4. ∎

Not only is Theorem 4.14 an interesting formulation of the separation theorem in itself, but it gives rise to the following question: If instead of being able to strictly separate every $n + 1$ points of B from A by a hyperplane, it is only possible to strictly separate every n points (or more generally every k points where $1 \leq k \leq n$) of B from A by a hyperplane, then is it possible in some way to "separate" all of B from A? The answer is yes. It turns out that there is a "cylinder" of an appropriate sort that contains A and that is disjoint from B. This topic will be pursued further in Chapter 4.

EXERCISES

4.1. Give an example of a compact set A and a closed set B in E^2 such that conv $A \cap$ conv $B = \emptyset$, but A and B cannot be strictly separated by a hyperplane.

4.2. Let S be a closed, convex proper subset of E^n. Prove that S is equal to the intersection of all the closed half-spaces that contain it.

4.3. Let $\pi\colon \mathsf{E}^n \to G$ be the projection mapping from E^n onto the $(n - k)$-dimensional subspace G spanned by the standard basis vectors $e_{k+1}, \ldots, e_n$ $(0 < k \leq n - 1)$. That is, $\pi(\alpha_1, \ldots, \alpha_n) \equiv (0, \ldots, 0, \alpha_{k+1}, \ldots, \alpha_n)$ for $(\alpha_1, \ldots, \alpha_n) \in \mathsf{E}^n$. Let S be an open convex subset of E^n. Prove that $\pi(S)$ is a relatively open convex subset of G.

4.4. Prove Theorem 4.14.

4.5. Prove that a closed half-space is a convex set.

4.6. Let A and B be disjoint open convex proper subsets of E^n. Prove that there exists a hyperplane strictly separating A and B.

4.7. In Figure 4.4, is α positive or negative? Why?

4.8. Let A and B be convex sets with int $B \neq \emptyset$. Prove that if $A \subset \operatorname{bd} B$ then aff $A \cap \operatorname{int} B = \emptyset$.

4.9. Let S be a proper, convex subset of E^n. Prove that cl S is also a proper convex subset of E^n.

4.10. (a) Prove: If A and B are nonempty disjoint convex subsets of E^n and $A \cup B$ is affine, then $A \cup B = \operatorname{aff} A = \operatorname{aff} B$.
(b) Show by examples that both of the conditions (disjoint and convex) are necessary in part (a).

4.11. Find a line L and a closed convex set S in E^3 such that $L \cap S = \emptyset$, but each plane containing L intersects S.

†4.12. Let F be a flat and let S be a convex subset of E^n with $F \cap S = \emptyset$. Find necessary and sufficient conditions for F to be contained in a hyperplane disjoint from S. (See Exercise 4.11.)

4.13. The hyperplane H is said to **strongly separate** two nonempty sets A_1 and A_2 if there exists some $\varepsilon > 0$ such that H strictly separates $A_1 + B(\theta, \varepsilon)$ and $A_2 + B(\theta, \varepsilon)$. [Recall that $B(\theta, \varepsilon)$ is the open ball of radius ε centered at θ.] Prove that two nonempty convex sets A_1 and A_2 can be strongly separated by a hyperplane iff $d(A_1, A_2) > 0$. (See Exercise 1.18.)

SECTION 5. SUPPORTING HYPERPLANES

We now turn our attention to one of the important applications of separation, which involves the existence of so-called "support" hyperplanes. This application can be used to characterize closed convex sets and to represent a compact convex set S as the convex hull of a "minimal" subset of S.

5.1. Definition. A hyperplane H is said to **support** a set S at a point $x \in S$ if $x \in H$ and if H bounds S.

5.2. Theorem. If x is a boundary point of a closed convex subset S of E^n, then there exists at least one hyperplane supporting S at x.

PROOF. If S is not n-dimensional, then any hyperplane containing S is a supporting hyperplane of S at each of the points of S.

If $\dim S = n$, then $\text{int } S \neq \emptyset$. Thus for any boundary point x, Corollary 4.6 implies that there exists a hyperplane H containing x that bounds $\text{int } S$. It follows that H also bounds S, and so H supports S at x. ∎

One of the surprising results in the study of convex sets is that a partial converse to Theorem 5.2 is also true.

5.3. Theorem. Let S be a closed subset of E^n such that $\text{int } S \neq \emptyset$. If through each boundary point of S there passes a hyperplane that supports S, then S is convex.

PROOF. We will prove the theorem for $n \geq 2$. The case for $n = 1$ is similar and is left as an exercise.

If $S = \mathsf{E}^n$, then clearly S is convex. Thus we may assume that there exists a point x not in S. (See Figure 5.1.) If $y \in \text{int } S$, then there exists a boundary point b of S in $\text{relint } \overline{xy}$. (See Exercise 2.11.) The supporting hyperplane H to S through b does not contain x, for otherwise it would contain the interior point y, which is not possible. Thus the closed half-space bounded by H and containing y includes S but does not contain x. Since x was any point not in S, we conclude that S is equal to the intersection of all the closed half-spaces that contain it. Thus S is an intersection of convex sets and so is convex itself. ∎

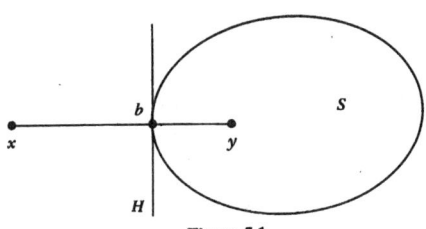

Figure 5.1.

Combining Theorems 5.2 and 5.3 we obtain the following characterization of closed convex sets with nonempty interior.

5.4. Theorem. Let S be a closed subset of E^n such that int $S \neq \emptyset$. Then S is convex iff through each boundary point of S there passes a hyperplane of support to S.

This theorem is truly remarkable in that it equates the global pairwise definition of convexity (the line segment joining *any two* points of S must lie in S) with a property of certain individual points (*each boundary* point must lie in a supporting hyperplane). Other characterizations of this sort are possible, and we shall return to them later (Chapter 7). An important application of supporting hyperplanes is presented now.

Given a convex set S, is it always possible to express S as the convex hull of some proper subset S^* of S? If S consists of more than one point, the answer is yes: Let $S^* = S \sim \{x\}$ where $x \in$ relint S. If S is compact, we may let $S^* =$ bd S. Is the boundary of a compact set the smallest subset with this property? A moment's reflection on a triangle or a square convinces one that in general the answer is no. The set of vertices would work just as well, and they are easily seen to be the smallest such set. Does there always exist a smallest subset S^* of S such that conv $S^* = S$? If it does exist, can it be characterized? In answering these questions, the following definition is useful.

5.5. Definition. Let S be a convex set. A point x in S is called an **extreme point** of S if there exists no nondegenerate line segment in S that contains x in its relative interior. The set of all extreme points of S is called the **profile** of S.

Examples. (See Figure 5.2.)

1. The profile of a polygon in E^2 is the set of vertices.
2. The profile of a closed ball is its boundary.
3. An open set has no extreme points, so its profile is empty.
4. A closed half-space has no extreme points, so its profile is empty.

It is easy to see that a point x in a convex set S is an extreme point of S iff the set $S \sim \{x\}$ is convex (see Exercise 5.1). It follows that any subset S^* of S

Figure 5.2.

SUPPORTING HYPERPLANES

such that conv $S^* = S$ must contain the profile of S. Examples 3 and 4 show that in general S^* may have to be larger than the profile of S. It is true, however, that when S is compact we may actually take S^* to be the profile of S (Theorem 5.6). Thus every compact set S has an extreme point, and the set of all extreme points is the smallest subset of S whose convex hull is equal to S.

For sets that are not compact, no simple characterization is possible. If S is not compact and $P \neq \emptyset$ is the profile of S, then either of the following may occur: conv $P = S$; or conv $P \neq S$. (See Exercises 5.2–5.4.)

5.6. Theorem. Let S be a compact convex subset of E^n. Then S is the convex hull of its profile.

PROOF. The proof is by induction on the dimension k of the set S. The cases $k = 0$ and $k = 1$ where S is a point or a closed line segment are trivial. Suppose the theorem is true for any compact convex set of dimension at most $k - 1$ where $k \leq n$. Let S have dimension k and let $x \in S$. If x is in the relative boundary of S, then applying Theorem 5.2 to S as a subset of aff S we obtain a hyperplane H (a flat of dimension $k - 1$) that supports S at x. The set $S \cap H$ is compact and convex, and its dimension does not exceed $k - 1$. By the induction hypothesis, x is a convex combination of the extreme points of $S \cap H$ and hence also of S (Exercise 5.5).

If x is in relint S, let L be a line in aff S through x. Then $L \cap S$ is a line segment with endpoints, say y and z, in the relative boundary of S. Since y and z are convex combinations of extreme points of S, so is x. Thus we have S contained in the convex hull of its profile. Since the reverse inclusion is always true, the two sets must be equal. ∎

There is a similarity between the profile of a compact convex set and a basis for a linear subspace that should be noted. A basis B for a linear subspace U is a linearly independent subset of U which (linearly) spans U in the sense that each element of U is a linear combination of elements of B. Each subspace U of E^n has a basis, and although the basis is not unique, each basis of U has the same finite number of elements in it. Similarly, if F is an affine subset of E^n, then there exists an affinely independent subset A of F having a finite number of elements such that aff $A = F$ (i.e., A affinely spans F.) If we define a set to be convexly independent if no element is a convex combination of the others, then for a compact convex set S, the profile P of S is a convexly independent subset of S such that conv $P = S$ (i.e., P convexly spans S.) Unfortunately, even though the profile is unique, it may have infinitely many members. Of course the analogy breaks down completely if the set S is not compact. This is one of the reasons that the theories of linear subspaces and affine sets are very similar, while the theory of convex sets has a uniqueness all its own.

This section concludes with an important application of Theorem 5.6 that will be useful in Chapter 10.

5.7. Theorem. Let f be a linear functional defined on the compact convex set S. Then there exist extreme points $\bar{x}$ and $\bar{y}$ of S such that

$$f(\bar{x}) = \max_{x \in S} f(x)$$

and

$$f(\bar{y}) = \min_{x \in S} f(x).$$

PROOF. By Theorem 3.4, f is continuous. Since S is compact, Theorem 1.21 implies the existence of a point x' in S such that

$$f(x') = \max_{x \in S} f(x) \equiv \alpha.$$

Again, the compactness of S together with Theorem 5.6 imply the existence of extreme points $x_1, \ldots, x_k$ of S and nonnegative $\lambda_1, \ldots, \lambda_k$ such that

$$x' = \sum_{i=1}^{k} \lambda_i x_i \quad \text{with} \quad \sum_{i=1}^{k} \lambda_i = 1.$$

If none of the extreme points of S satisfy $f(x) = \alpha$, then we must have

$$f(x_i) < \alpha \quad \text{for } i = 1, \ldots, k$$

since α is the maximum of f on S. But this means that

$$\alpha = f(x') = \sum_{i=1}^{k} \lambda_i f(x_i)$$

$$< \sum_{i=1}^{k} \lambda_i \alpha = \alpha \sum_{i=1}^{k} \lambda_i = \alpha.$$

This contradiction implies that some extreme point $\bar{x}$ of S must satisfy $f(\bar{x}) = \alpha$.

The proof for $\bar{y}$ is completely analogous. ∎

EXERCISES

5.1. Let x be an element of the convex set S. Prove that x is an extreme point of S if and only if $S \sim \{x\}$ is convex.

5.2. Find an example of an unbounded convex set in E^2 that is equal to the convex hull of its profile.

5.3. Find an example of a closed convex set S in E^2 such that its profile P is nonempty but $\operatorname{conv} P \neq S$.

EXERCISES

5.4. Find an example of a bounded convex set S in E^2 such that its profile P is nonempty but $\operatorname{conv} P \neq S$.

5.5. Let S be a closed convex subset of E^n and suppose H is a supporting hyperplane to S at the point x. Prove that x is an extreme point of $H \cap S$, iff x is an extreme point of S.

5.6. Find an example of a compact convex subset S of E^3 such that its profile P is not closed.

5.7. Find an example of a closed nonconvex subset of E^2 with the property that through each of its boundary points there passes a hyperplane of support. (Compare with Theorem 5.3.)

5.8. Prove the following: Let A and B be nonempty compact subsets of E^2 with $x_1 \in A$ and $x_2 \in B$. If $H_1 \equiv [f: \alpha_1]$ supports A at x_1 with $f(A) \geq \alpha_1$ and $H_2 \equiv [f: \alpha_2]$ supports B at x_2 with $f(B) \geq \alpha_2$, then $H_3 \equiv [f: \alpha_1 + \alpha_2]$ supports $A + B$ at $x_1 + x_2$ with $f(A + B) \geq \alpha_1 + \alpha_2$. (Note that H_1, H_2, and H_3 are all parallel since they are described by the same linear functional f.)

5.9. A point x in the boundary of a closed convex set S is called an **exposed point** of S if there exists a hyperplane of support H to S through x for which $H \cap S = \{x\}$.
   (a) Prove or disprove: Each exposed point of S is an extreme point of S.
   (b) Prove or disprove: Each extreme point of S is an exposed point of S.

5.10. Prove Theorem 5.3 for the case $n = 1$.

5.11. Let S^* be a subset of the convex set S such that $\operatorname{conv} S^* = S$. Prove that S^* contains the profile of S.

5.12. (a) Let S be a bounded convex set. Prove that $\operatorname{conv}(\operatorname{bd} S) = \operatorname{cl} S$.
   (b) Find an example to show that the boundedness of S is necessary in part (a).

5.13. Let S be a compact convex set with $\operatorname{int} S \neq \emptyset$. Use Theorem 5.6 to prove that the boundary of S is not convex.

5.14. Prove the contrapositive of Theorem 5.3 directly without using Theorem 5.3. That is, suppose S is closed and $\operatorname{int} S \neq \emptyset$. If S is not convex, find a boundary point of S that lies on no hyperplane of support to S.

5.15. A convex subset C of E^n that has at least two points is called a **convex cone** with vertex x_0 if for each $\lambda \geq 0$ and for each $x \in C$, $x \neq x_0$, we have
$$(1 - \lambda)x_0 + \lambda x \in C.$$

A cone that is a proper subset of a line is called a **ray** or **half-line**. Prove the following:
   (a) Every closed convex cone C that is a proper subset of E^n has at least one hyperplane of support, and each hyperplane of support contains its vertex x_0.

(b) The closure of a convex cone with vertex x_0 is a convex cone with vertex x_0.

(c) Let C_α ($\alpha \in \mathcal{Q}$) be a family of convex cones each having vertex x_0. If $C \equiv \bigcap_{\alpha \in \mathcal{Q}} C_\alpha$ contains at least two points, then C is a convex cone with vertex x_0.

5.16. Let x_0 and p be distinct points, and let $R(x_0, p)$ denote the closed ray with vertex x_0 that contains the point p. Let C be a convex cone having vertex x_0. Prove the following:

(a) If $q \in R(x_0, p)$ and $q \neq x_0$, then $R(x_0, p) = R(x_0, q)$.

(b) If $C \neq \mathsf{E}^n$, then $x_0 \in \mathrm{bd}\, C$.

(c) If $y \in \mathrm{bd}\, C$ and $y \neq x_0$, then $R(x_0, y) \subset \mathrm{bd}\, C$.

(d) If $y \in \mathrm{int}\, C$, then $\mathrm{relint}\, R(x_0, y) \subset \mathrm{int}\, C$.

5.17. Let S be a closed, convex, proper subset of E^n containing at least two points. Then S is a convex cone iff there exists at least one point common to all the hyperplanes that support S.

5.18. Let S be a nonempty compact set and let pos S denote its positive hull. (See Exercise 2.24.) Prove that pos S is a closed convex cone with vertex at θ.

3

HELLY-TYPE THEOREMS

At the end of Section 4 a question that is representative of a whole class of theorems was raised. In general, these results can be written in the following form:

Let $\mathscr{F}$ be a collection of sets and let k be a fixed integer. If every k sets in $\mathscr{F}$ have property P, then the collection $\mathscr{F}$ has property Q.

One of the earliest and most widely known theorems of this type is due to Edward Helly (1884–1943). It was discovered by him and communicated to J. Radon in 1913, just prior to Helly's joining the Austrian army. During the war Helly was wounded by the Russians and taken as a prisoner to Siberia. He returned to Vienna in 1920 and the following year received an appointment at the University of Vienna. The first proof of his theorem was published by Radon in 1921 and his own proof appeared two years later. In 1938 Helly and his family emigrated to the United States.

SECTION 6. HELLY'S THEOREM

Since the 1920s a variety of proofs of Helly's theorem (Theorem 6.2) have appeared. Some are geometric, some are algebraic, and others use duality (see Chapter 9). We will follow the approach used by Radon, and begin with a basic theorem that bears his name.

6.1. Theorem (Radon's Theorem). Let $S \equiv \{x_1, x_2, \ldots, x_r\}$ be any finite set of points in E^n. If $r \geq n + 2$, then S can be partitioned into two disjoint subsets S_1 and S_2 such that $\operatorname{conv} S_1 \cap \operatorname{conv} S_2 \neq \varnothing$.

PROOF. Since $r \geq n + 2$, Theorem 2.18 implies that S is affinely dependent. Thus there exist scalars $\alpha_1, \ldots, \alpha_r$, not all zero, such that

$$\sum_{i=1}^{r} \alpha_i x_i = \theta \quad \text{and} \quad \sum_{i=1}^{r} \alpha_i = 0.$$

47

At least two of the α_i's must have opposite signs, so we may assume without loss of generality that for some k $(1 \leq k < r)$,

$$\alpha_1 \geq 0, \ldots, \alpha_k \geq 0, \alpha_{k+1} < 0, \ldots, \alpha_r < 0.$$

Since the α_i's sum to zero, we may let

$$\alpha \equiv \alpha_1 + \cdots + \alpha_k = -(\alpha_{k+1} + \cdots + \alpha_r).$$

Then $\alpha > 0$ and we can let

$$x \equiv \sum_{i=1}^{k} \frac{\alpha_i}{\alpha} x_i = \sum_{i=k+1}^{r} \left(-\frac{\alpha_i}{\alpha}\right) x_i.$$

It follows that x is a convex combination of $x_1, \ldots, x_k$, and so by Theorem 2.22 $x \in \text{conv}\{x_1, \ldots, x_k\}$. Likewise, $x \in \text{conv}\{x_{k+1}, \ldots, x_r\}$. Thus letting $S_1 \equiv \{x_1, \ldots, x_k\}$ and $S_2 \equiv \{x_{k+1}, \ldots, x_r\}$, we have $\text{conv}\,S_1 \cap \text{conv}\,S_2 \neq \varnothing$. ∎

6.2. Theorem (Helly's Theorem). Let $\mathcal{F} \equiv \{B_1, \ldots, B_r\}$ be a family of r convex sets in E^n with $r \geq n + 1$. If every subfamily of $n + 1$ sets in $\mathcal{F}$ has a nonempty intersection, then $\bigcap_{i=1}^{r} B_i \neq \varnothing$.

PROOF. The proof is by induction on the number r of convex sets in the collection. If $r = n + 1$, then the theorem is trivially true. Suppose now that the theorem holds for every collection of $r - 1$ convex sets. This implies that for each i $(1 \leq i \leq r)$ there exists a point x_i such that

$$x_i \in B_1 \cap \cdots \cap B_{i-1} \cap B_{i+1} \cap \cdots \cap B_r.$$

Since $r \geq n + 2$ we can apply Radon's Theorem to the set $S = \{x_1, \ldots, x_r\}$ and obtain two disjoint subsets S_1 and S_2 of S such that $\text{conv}\,S_1 \cap \text{conv}\,S_2 \neq \varnothing$. Without loss of generality we may let $S_1 \equiv \{x_1, \ldots, x_k\}$ and $S_2 \equiv \{x_{k+1}, \ldots, x_r\}$, where $1 \leq k < r$. We now choose

$$x \in \text{conv}\{x_1, \ldots, x_k\} \cap \text{conv}\{x_{k+1}, \ldots, x_r\}$$

and claim that x is in each B_i, $i = 1, \ldots, r$. Indeed, for each $i \leq k$ we have $x_i \in B_{k+1} \cap \cdots \cap B_r$. Therefore, since each B_i is convex,

$$x \in \text{conv}\{x_1, \ldots, x_k\} \subset B_{k+1} \cap \cdots \cap B_r.$$

Similarly, when $i \geq k + 1$ we have $x_i \in B_1 \cap \cdots \cap B_k$, so $x \in \text{conv}\{x_{k+1}, \ldots, x_r\} \subset B_1 \cap \cdots \cap B_k$. ∎

The proof of Helly's Theorem is illustrated in Figure 6.1 for four sets in the plane. In this case $S_1 = \{x_1, x_2\}$ and $S_2 = \{x_3, x_4\}$.

HELLY'S THEOREM

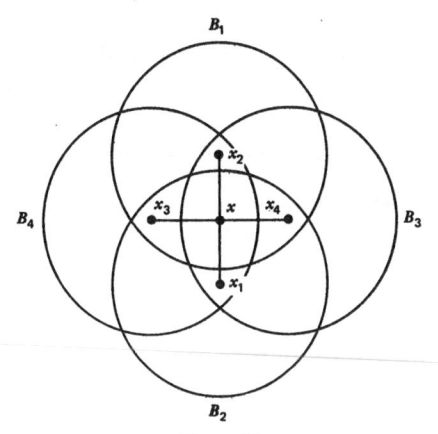

Figure 6.1.

The various generalizations, adaptations, and applications of Helly's Theorem seem to be endless. Several of the more interesting ones will be presented in this chapter, and others will be scattered throughout the remaining chapters.

It is easy to see that the sets in Helly's Theorem must all be convex. For example, every three of the sets in Figure 6.2 have a nonempty intersection, but there is no point in common to all four sets. It also follows readily that in general the "Helly number" $n + 1$ cannot be reduced. For example, every two sides of a triangle in E^2 have a point in common, but all three sides do not.

Helly's Theorem can be extended to infinite collections of convex sets, but not without some additional restrictions. (See Exercises 6.2 and 6.3.) We may require, for example, that all the sets be compact.

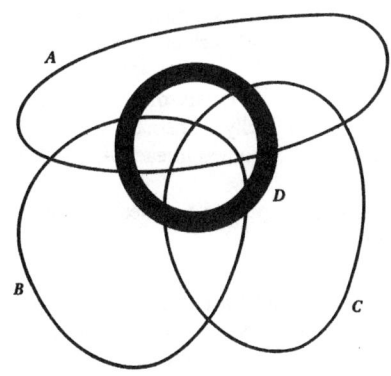

Figure 6.2.

6.3. Theorem (Helly's Theorem). Let $\mathcal{F}$ be a family of compact convex subsets of E^n containing at least $n + 1$ members. If every $n + 1$ members of $\mathcal{F}$ have a point in common, then all the members of $\mathcal{F}$ have a point in common.

PROOF. This is proved in Section 24 (Theorem 24.9) using duality, but it can also be proved directly (see Exercise 6.4). ∎

In considering Helly's Theorem, it is natural to ask if one can conclude anything about the "size" of the intersections involved. This question was answered in the affirmative by Victor Klee in 1953 when he proved the following theorem.

6.4. Theorem. Let $\mathcal{F} \equiv \{A_\alpha: \alpha \in \mathcal{Q}\}$ be a family of compact convex subsets of E^n containing at least $n + 1$ members. Suppose K is a compact convex subset of E^n such that the following holds: For each subfamily of $n + 1$ sets in $\mathcal{F}$, there exists a translate of K that is contained in all $n + 1$ of them. Then there exists a translate of K that is contained in all the members of $\mathcal{F}$.

PROOF. For each A_α in $\mathcal{F}$, let

$$A_\alpha^* \equiv \{p: (p + K) \subset A_\alpha\}.$$

We claim that each A_α^* is convex. To see this let x and y be elements of A_α^* and let $0 \leq \lambda \leq 1$. We must show that

$$[\lambda x + (1 - \lambda)y] + K \subset A_\alpha.$$

But for each $k \in K$ we have

$$\lambda x + (1 - \lambda)y + k = \lambda(x + k) + (1 - \lambda)(y + k).$$

Since $x + k \in A_\alpha$ and $y + k \in A_\alpha$ and since A_α is convex, the right-hand side of the preceding equation is also in A_α. Hence $\lambda x + (1 - \lambda)y \in A_\alpha^*$ and A_α^* is convex. Furthermore, since K and each A_α are compact, each A_α^* is compact also.

By hypothesis, every $n + 1$ members of the family $\{A_\alpha^*: \alpha \in \mathcal{Q}\}$ have a point in common. Hence, by Helly's Theorem 6.3, there exists a point q in $\cap_{\alpha \in \mathcal{Q}} A_\alpha^*$. But then $q + K$ is contained in each A_α ($\alpha \in \mathcal{Q}$). ∎

The following theorem (also proved by Klee) is similar to Theorem 6.4 and will be useful to us in Chapter 5.

6.5. Theorem. Let $\mathcal{F}$ be a family of compact convex subsets of E^n containing at least $n + 1$ members. Suppose K is a compact convex subset of E^n such that for each subfamily of $n + 1$ sets in $\mathcal{F}$, there exists a translate of K that contains all $n + 1$ of them. Then there exists a translate of K that contains all the members of $\mathcal{F}$.

HELLY'S THEOREM

PROOF. The proof is similar to the proof of Theorem 6.4 and is left to the reader (Exercise 6.5). ∎

Another way to extend Helly's Theorem is to require that each subfamily of k sets have a common point, where $1 \leq k \leq n$. If we weaken this intersection condition, we of course expect a weaker conclusion. The case when $k = n$ is presented in the following theorem, and the general case is in the subsequent corollary. Both results are originally due to A. Horn (1949), and the proofs presented here follow those of Eggleston (1966).

6.6. Theorem. Let $\mathscr{F} \equiv \{B_1, \ldots, B_r\}$ be a family of r compact convex subsets of E^n with $r \geq n$. If every subfamily of n members of $\mathscr{F}$ has a point in common, then given any point y in E^n there exists a line through y that intersects all the members of $\mathscr{F}$.

PROOF. The proof is by induction on the number r of sets. The result is trivial if $r = n$. Suppose inductively that the theorem is true for every collection of $r - 1$ compact convex sets. We assume without loss of generality that $y = \theta$ and that y does not belong to any of the sets in $\mathscr{F}$. Indeed, if y were in one of the sets in $\mathscr{F}$, then the result would follow immediately by applying the induction hypothesis to the remaining $r - 1$ sets in $\mathscr{F}$.

Let S be the surface of an n-dimensional ball centered at y (the origin). For each $i = 1, \ldots, r$ we define $B_i^* \equiv S \cap (\text{pos } B_i)$. Then each set B_i^* is contained in an open hemisphere on S. By the induction hypothesis there exists a point

$$w_j \in \bigcap_{i \neq j} B_i^*, \quad \text{for each } j = 1, 2, \ldots, r.$$

Now either w_1 and B_1^* lie in an open hemisphere on S or the line through w_1 and y meets B_1^*. In the latter case, the theorem is proved. In the former case, there exists a hyperplane H through y such that w_1 and B_1^* lie on the same side of H. Since $w_j \in B_1^*$ for all $j \neq 1$, it follows that all the points w_j ($j = 1, \ldots, r$) lie on the same side of H.

On the line through y and w_j choose a point $x_{j,i}$ in B_i, $i \neq j$. For $i = 1, \ldots, r$ let $A_i \equiv \text{conv} \bigcup_{j \neq i} x_{j,i}$. Then $A_i \subset B_i$. Now let π be the radial projection from y onto a hyperplane H' parallel to H. (See Figure 6.3.) It follows that the sets $\pi(A_i)$ form a collection of r compact convex sets in H' each n of which have a point in common. Since $\dim H' = n - 1$ we may apply Helly's Theorem 6.2 to obtain a point in common to all these projections. It follows that the line through this point and y intersects all the sets B_i ($i = 1, \ldots, r$). ∎

6.7. Corollary. Let $\mathscr{F}$ be a family of r compact convex subsets of E^n with $r \geq n$. Suppose that every subfamily of k members of $\mathscr{F}$ have a point in common, where $1 \leq k \leq n$. Then given any $(n - k)$-dimensional flat F_1, there exists a $(n - k + 1)$-dimensional flat F_2 such that $F_2 \supset F_1$ and F_2 intersects all the members of $\mathscr{F}$.

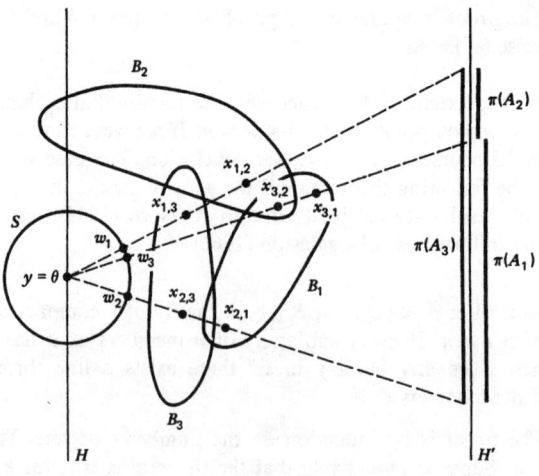

Figure 6.3.

PROOF. The case when $k = n$ is Theorem 6.6, so we may assume that $1 \leq k < n$. Let U be the k-dimensional subspace orthogonal to F_1. Project E^n onto U by means of $(n - k)$-dimensional flats parallel to F_1. Then in U we have a family of compact convex sets every k of which have a common point. Since F_1 projects onto a point, say y, and $\dim U = k$, we may apply Theorem 6.6 to obtain a line through y in U that meets the projection of each set in $\mathcal{F}$. This line is the projection of a $(n - k + 1)$-dimensional flat F_2 that contains F_1 and that intersects every set in $\mathcal{F}$. ∎

It should be noted that Corollary 6.7 can be extended to infinite families since the sets are compact. In fact, this is the form in which the theorem was originally proved by Horn. The proof for the infinite case is more difficult, however, and will not be presented here.

6.8. Theorem (Horn). Let $\mathcal{F}$ be a family of compact convex subsets of E^n containing at least n members. Suppose that every subfamily of k members of $\mathcal{F}$ have a point in common, where $1 \leq k \leq n$. Then given any $(n - k)$-dimensional flat F_1, there exists a $(n - k + 1)$-dimensional flat F_2 such that $F_2 \supset F_1$ and F_2 intersects all the members of $\mathcal{F}$.

One of the interesting applications of Helly's Theorem is the "art gallery theorem" of Krasnosselsky (1946). It can be described colloquially as follows. Imagine an art gallery consisting of several connected rooms in which the walls are completely covered with pictures. The theorem implies that if for each three paintings in the gallery there exists a point from which all three can be seen, then there exists a point from which *all* the paintings in the gallery can be seen.

More precisely stated, Krasnosselsky's Theorem gives a Helly-type combinatorial criterion for determining whether or not a compact set is star-shaped. The theorem is preceded by a definition.

6.9. Definition. Let S be a subset of E^n. A point y in S is said to be **visible** from a point x in S if $\overline{xy} \subset S$.

6.10. Theorem (Krasnosselsky). Let S be a compact subset of E^n that contains at least $n+1$ points. Suppose that for each $n+1$ points of S there is a point from which all $n+1$ are visible. Then the set S is star-shaped.

PROOF. For each $x \in S$, let $V_x \equiv \{y: \overline{xy} \subset S\}$. The hypotheses imply that each $n+1$ of the sets V_x have a common point, and we must show that they all do. By Helly's Theorem 6.3 there exists a point $y \in \bigcap_{x \in S} \text{conv } V_x$, and we claim that $y \in \bigcap_{x \in S} V_x$.

Suppose the contrary: $y \notin \bigcap_{x \in S} V_x$. Then there exist $x \in S$ and $u \in \overline{xy}$ such that $u \notin S$. Let x' be a point of the relative boundary of S in $\overline{xu}$. (See Figure 6.4.) Since S is compact, there exists a point w in $\overline{x'u}$ such that

$$d(w, x') = \tfrac{1}{2} d(u, S).$$

Furthermore, there exist points v in $\overline{wu}$ and z in S such that

$$d(v, z) = d(\overline{wu}, S).$$

Since z is a point of S nearest to v, it follows that V_z lies in the closed half-space P that misses v and is bounded by the hyperplane H through z orthogonal to $\overline{vz}$. But since $y \in \text{conv } V_z \subset P$, $\angle yzv \geq \pi/2$ and so $\angle zvy < \pi/2$. Furthermore, since $d(v, S) \leq d(w, S) < d(u, S)$, it follows that $v \neq u$. Thus some point of $\overline{vu}$ is closer to z than v is. This contradicts the choice of v and completes the proof. ■

EXERCISES

6.1. Find an example of four points in the plane to show that the partitioning in Radon's Theorem 6.1 is not unique.

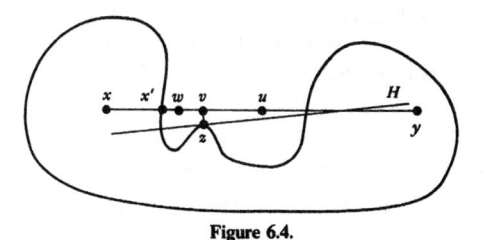

Figure 6.4.

6.2. Find a counterexample to Helly's Theorem applied to infinitely many convex sets that are bounded but not closed.

6.3. Find a counterexample to Helly's Theorem applied to infinitely many convex sets that are closed but not bounded.

*6.4. Prove Theorem 6.3.

6.5. Prove Theorem 6.5.

6.6. Let $\mathcal{F}$ be a family of compact convex subsets of E^n containing at least $n + 1$ members. Suppose that for each subfamily of $n + 1$ members of $\mathcal{F}$ there exists a point whose distance from each of the $n + 1$ convex sets is less than or equal to a fixed positive number d. Prove that there exists a point whose distance from each set in $\mathcal{F}$ is less than or equal to d.

6.7. Consider the following points in the plane: $x_1 = (0,0)$, $x_2 = (0,2)$, $x_3 = (2,3)$, $x_4 = (3,2)$, and $x_5 = (2,0)$. Use the technique in the proof of Radon's Theorem 6.1 to find a point x in $\text{conv}\{x_1, x_3, x_4\} \cap \text{conv}\{x_2, x_5\}$.

6.8. Let S be the set consisting of the following six points in E^3: $x_1 = (2,0,-1)$, $x_2 = (1,1,2)$, $x_3 = (0,-1,1)$, $x_4 = (-1,0,0)$, $x_5 = (1,0,1)$, $x_6 = (0,-3,3)$. Observe that

$$x_1 + x_2 - 2x_3 + x_4 - 2x_5 + x_6 = (0,0,0).$$

Find a partitioning of S into two disjoint subsets S_1 and S_2 such that $\text{conv}\,S_1 \cap \text{conv}\,S_2 \neq \emptyset$, and find the coordinates of a point y in $\text{conv}\,S_1 \cap \text{conv}\,S_2$.

6.9. Let $\{a, b, c, d\}$ be a subset of E^2 such that $\text{conv}\{a, b, c\} \cap \text{conv}\{a, b, d\} \cap \text{conv}\{a, c, d\} = \{a\}$. Prove that $a \in \text{conv}\{b, c, d\}$.

6.10. State and prove a generalization of Exercise 6.9 for E^n.

6.11. Let $\mathcal{F}$ be a family of compact convex subsets of E^2 containing at least two members. Suppose that each two members of $\mathcal{F}$ have a point in common. If L is an arbitrary line in the plane, show that L or a line parallel to L intersects all the sets in $\mathcal{F}$.

6.12. State and prove a generalization of Exercise 6.11 to E^n.

6.13. Prove the following converse to Radon's theorem: Let $S \equiv \{x_1, \ldots, x_r\}$ be a finite subset of E^n with $2 \leq r \leq n + 1$. If S is affinely independent, then for any partition of S into nonempty disjoint subsets S_1 and S_2, we must have $\text{conv}\,S_1 \cap \text{conv}\,S_2 = \emptyset$.

*6.14. Prove Theorem 6.8.

*6.15. Use Helly's Theorem 6.2 to prove Radon's Theorem 6.1.

*6.16. Use Krasnosselsky's Theorem 6.10 to prove Helly's Theorem 6.2.

*6.17. Let C be a compact convex subset of E^2 and let $S \subset \mathsf{E}^2 \sim C$. A point x in S is said to **see a point** $p \in C$ **via the complement** of C if

$\overline{xp} \cap C = \{p\}$. Prove the following:
- (a) If every five or fewer points in S can see some point of C via the complement of C, then there exists at least one point $p \in C$ that all the points of S can see via the complement of C.
- (b) If C contains exactly four, three, or two vertices, then part (a) holds with the word *five* replaced by *four*. (See Definition 11.5 for vertex.)
- (c) If C contains at most one vertex, then part (a) holds with the word *five* replaced by *three*.

*6.18. Let K be a compact subset of E^2 with int $K \neq \emptyset$, and let S be a closed subset of bd K. If $x \in S$ and $p \in \mathsf{E}^2$, then we say that x can see p externally relative to K if $\overline{xp} \cap K = \{x\}$. Prove the following: If every five points of S can see some common point externally relative to K, then all of S can see some common point externally relative to K.

*6.19. Let S be a compact subset of E^n. Suppose that every $n+1$ points of S can see via S a common n-dimensional ball of radius $\varepsilon > 0$. (That is, if x is one of the $n + 1$ points and y is in the ball, then $\overline{xy} \subset S$.) Prove that the kernel of S contains an n-dimensional ball of radius ε.

6.20. Let $\mathcal{F}$ be a finite family of subsets of E^2. We say that $\mathcal{F}$ has property $T(m)$ if every m-membered subfamily of $\mathcal{F}$ has a common transversal. That is, there exists a line that intersects each of the sets in the subfamily. We say that $\mathcal{F}$ has property $T - k$ if there exists a line intersecting all but k or fewer members of $\mathcal{F}$. We say that $\mathcal{F}$ has property T if there exists a line intersecting each member of $\mathcal{F}$.
- (a) Let $\mathcal{F}$ consist of three or more parallel line segments. Prove that if $\mathcal{F}$ has property $T(3)$, then $\mathcal{F}$ has property T.
- *(b) Let $\mathcal{F}$ consist of five or more pairwise disjoint translates of a parallelogram. Prove that if $\mathcal{F}$ has property $T(5)$, then $\mathcal{F}$ has property T.
- *(c) Prove that there exists a positive integer k such that for any finite family $\mathcal{F}$ of pairwise disjoint translates of a compact convex set, property $T(3)$ implies property $T - k$.
- †(d) What is the smallest k that will work in part (c)?
- †(e) Does there exist a positive integer m such that for any finite family $\mathcal{F}$ of pairwise disjoint translates of a compact convex set, property $T(m)$ implies property T?

SECTION 7. KIRCHBERGER'S THEOREM

Suppose a valley contains a number of stationary sheep and goats. What simple condition will guarantee that a straight fence can be built which will separate the sheep from the goats? The answer is perhaps surprising. If every four animals can be separated in this manner, then all the sheep can be separated from all the goats. We will not concern ourselves with the theological implications of this fact, but merely note that the formal statement of this result was first established in 1902 by Paul Kirchberger.

7.1. Theorem (Kirchberger). Let P and Q be nonempty compact subsets of E^n. Then P and Q can be strictly separated by a hyperplane iff for each set T consisting of $n + 2$ or fewer points from $P \cup Q$ there exists a hyperplane that strictly separates $T \cap P$ and $T \cap Q$.

It is indeed remarkable that Kirchberger was able to prove this at such an early date, especially since it antedated Helly's celebrated theorem by 20 years and even the basic theorem of Caratheodory by 5 years. As might be expected, Kirchberger's original proof was quite long—over 20 pages. Since then several shorter proofs have been given, including those of Rademacher and Schoenberg in 1950 and Shimrat in 1955. The proof given here uses the existence of a certain extreme point to shorten the argument even more. It is preceded by a definition and a lemma.

7.2. Definition. A point p in E^n has the **k-point simplicial property** with respect to a set $S \subset E^n$ if there exists a simplex $\Delta \equiv \text{conv}\{x_1, x_2, \ldots, x_r\}$ with $r \leq k$ that has vertices in S and that contains p.

For the sake of brevity in the following theorem, we shall for any set S let S_k ($k = 1, \ldots, n + 1$) be the set of all points in E^n that have the k-point simplicial property with respect to S. It is clear that $S_1 = S$, and by Caratheodory's Theorem 2.23 we have $S_{n+1} = \text{conv} S$.

7.3. Lemma. Suppose A and B are two compact subsets of E^n such that $\text{conv} A \cap \text{conv} B \neq \varnothing$. Let x be an extreme point of $\text{conv} A \cap \text{conv} B$ and let i and j be the smallest integers such that $x \in A_i \cap B_j$. Then $i + j \leq n + 2$.

PROOF. Since the convex hull of a compact set is compact, it follows from Theorem 5.6 that there does in fact exist an extreme point x of $\text{conv} A \cap \text{conv} B$. Now since i is the smallest integer such that $x \in A_i$, x is the center of a $(i - 1)$-dimensional ball Γ_A of positive radius with $\Gamma_A \subset \text{conv} A$. Since j is the smallest integer such that $x \in B_j$, x is the center of a $(j - 1)$-dimensional ball Γ_B of positive radius with $\Gamma_B \subset \text{conv} B$. Now $\Gamma_A \cap \Gamma_B$ must contain only x since x is an extreme point of $\text{conv} A \cap \text{conv} B$. It follows that $(i - 1) + (j - 1) \leq n$, that is, $i + j \leq n + 2$. ∎

PROOF OF THEOREM 7.1. Suppose that for each subset T of $n + 2$ or fewer points of $P \cup Q$ there exists a hyperplane that strictly separates $T \cap P$ from $T \cap Q$, and suppose that P and Q cannot be strictly separated by a hyperplane. Then by Theorem 4.13 $\text{conv} P \cap \text{conv} Q \neq \varnothing$, and we may choose an extreme point, say x, of $\text{conv} P \cap \text{conv} Q$. By Lemma 7.3 there exist points

KIRCHBERGER'S THEOREM

$p_1, \ldots, p_i$ in P and $q_1, \ldots, q_j$ in Q with $i + j \leq n + 2$ such that x is in

$$[\text{conv}\{p_1, \ldots, p_i\}] \cap [\text{conv}\{q_1, \ldots, q_j\}].$$

This contradicts the assumption that there exists a hyperplane strictly separating $\{p_1, \ldots, p_i\}$ and $\{q_1, \ldots, q_j\}$.

The converse is immediate. ∎

It is easy to see that Kirchberger's Theorem is a "best" theorem in the sense that the number $n + 2$ cannot be reduced. For example, let P consist of the vertices of an n-dimensional simplex in E^n. Let $Q \equiv \{q\}$ be a point in the interior of the convex hull of P. Since q is not in the convex hull of any n points of P, given any subset T of $n + 1$ or fewer points of $P \cup Q$, there exists a hyperplane strictly separating $T \cap P$ and $T \cap Q$. But since $q \in \text{conv}\, P$, it follows that Q and P cannot be strictly separated by a hyperplane.

Just as Klee's extension of Helly's Theorem determined the size of the intersection involved, it is possible to extend Kirchberger's Theorem to determine the size of the separation. The following definitions are useful in this regard.

7.4. Definition. A **slab** in E^n is the closed connected region bounded by two distinct parallel hyperplanes. The width of a slab is the distance between its bounding hyperplanes.

7.5. Definition. Suppose that f is a nonzero linear functional defined on E^n and that α and β are real numbers with $\alpha < \beta$. The slab

$$S \equiv \{x : \alpha \leq f(x) \leq \beta\}$$

is said to **strictly separate** two subsets P and Q of E^n if either $f(P) > \beta$, $f(Q) < \alpha$, or $f(P) < \alpha$, $f(Q) > \beta$ holds.

7.6. Theorem. Let P and Q be nonempty compact subsets of E^n. Then P and Q can be strictly separated by a slab of width $\delta > 0$ if and only if for each subset T of $n + 2$ or fewer points of $P \cup Q$ there exists a slab of width δ that strictly separates $T \cap P$ and $T \cap Q$.

PROOF. Exercise 7.3. ∎

7.7. Application. Let $\Phi(x, y)$ be a function of the two independent variables x and y, and suppose that the domain $D \equiv \{(x_i, y_i) : i = 1, \ldots, r\}$ of Φ is a finite subset of E^2. There are many ways to approximate the function Φ over its domain D. One way is to require that the *largest error* that would be made by replacing Φ by its approximation should be *as small as possible*. Let us consider the problem of finding a linear function $f(x, y) \equiv ax + by + c$ that approximates Φ in this way.*

Geometrically, we are given r points in three-space

$$(x_1, y_1, z_1), (x_2, y_2, z_2), \ldots, (x_r, y_r, z_r)$$

where $z_i = \Phi(x_i, y_i)$, and we are looking for a plane

$$z = f(x, y) = ax + by + c$$

such that the largest of the differences $|f(x_i, y_i) - z_i|$ is made as small as possible. For a given approximating function f and its corresponding plane H, we define the **deviation** δ by

$$\delta \equiv \max\{|f(x_i, y_i) - z_i| : i = 1, \ldots, r\}.$$

There may, of course, be several points at which the deviation is attained. Some may represent values below the plane (i.e., where the approximation is too large), and others represent values above the plane (i.e., where the approximation is too small). Let

$$P \equiv \{(x_i, y_i) : f(x_i, y_i) - \delta = z_i\}$$

and

$$Q \equiv \{(x_i, y_i) : f(x_i, y_i) + \delta = z_i\}.$$

Then a necessary and sufficient condition that f be the "best" linear approximation to Φ is that $\operatorname{conv} P \cap \operatorname{conv} Q \neq \varnothing$.

Indeed, suppose $\operatorname{conv} P \cap \operatorname{conv} Q = \varnothing$. Then there exists a line L in the xy-plane that strictly separates P and Q. Let L_1 be a line in the plane $z = f(x, y)$ directly above L. (See Figure 7.1.) A small rotation of the approximating plane about the line L_1 will reduce the difference $f(x_i, y_i) - z_i$ at all P points (where it is positive) and increase it at all Q points (where it is negative). Such a rotation may in fact increase $|f(x_i, y_i) - z_i|$ at other points of the domain, but since the differences at these points are all *less* than δ, a sufficiently small rotation will keep them less than δ. Thus if $\operatorname{conv} P \cap \operatorname{conv} Q = \varnothing$, we can find a slight rotation of the plane that will yield a smaller deviation.

On the other hand, if there exists a better approximation than f, then the corresponding plane H' is lower than H at each point of $\operatorname{conv} P$ and higher than H at each point of $\operatorname{conv} Q$. This implies $\operatorname{conv} P \cap \operatorname{conv} Q = \varnothing$. Thus we conclude that f is the best approximation to Φ iff $\operatorname{conv} P \cap \operatorname{conv} Q \neq \varnothing$.

---

*We are using the term "linear" in a wider sense here than in Definition 3.1. Technically, the function f used here should be called "affine," but the preceding terminology is standard.

Figure 7.1.

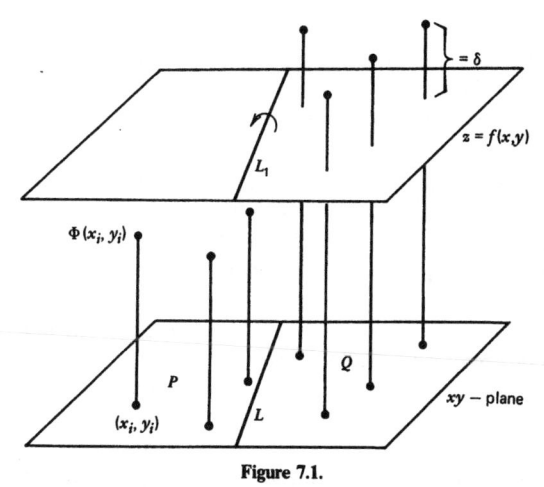

From Kirchberger's Theorem 7.1, we know that $\text{conv } P \cap \text{conv } Q \neq \emptyset$ iff there exist four points in $P \cup Q$ such that the convex hull of the p's intersects the convex hull of the q's. Thus if we are looking for the best approximating plane to a given set of points, there are four points from among them such that the best approximating plane of these four points is also the best approximating plane of all the points.

This then suggests a method for finding the linear function f that best approximates Φ on the domain D. For each subset T consisting of four points of D, find the best linear approximation to these four points and note the deviation δ_T. (In fact, T can be chosen from among the points in D that correspond to extreme points of the three-dimensional set.) Among all these deviations let δ be the largest. The corresponding linear function will be the best approximation to Φ over the whole domain D.

The preceding discussion can, of course, be generalized to a function Φ of n variables. (An example for $n = 1$ may be found in Exercise 7.2.) In this case we determine the best linear approximation for subsets T of $n + 2$ points of D and again select the one with the largest deviation. It will be the best approximation over the whole domain.

EXERCISES

7.1. Let S be the vertices of a square in E^2. Let S_k ($k = 1, 2, 3$) be the set of all points in E^2 that have the k-point simplicial property with respect to S. Describe the sets S_1, S_2, and S_3.

7.2. Consider the following four points in E^2: $x_1 = (1, 1)$, $x_2 = (2, 3)$, $x_3 = (3, 2)$, and $x_4 = (4, 3)$. Use the techniques of Application 7.7 to find the

following:
(a) The deviation for the best linear approximation to each subset of three points.
(b) The best linear approximation to all four points.

7.3. Prove Theorem 7.6.

*7.4. Prove the following Kirchberger-type theorem for (weak) separation: Let P and Q be n-dimensional subsets of E^n. Then P and Q can be separated by a hyperplane iff for each subset T of $2n + 2$ or fewer points of $P \cup Q$ there exists a hyperplane that separates $T \cap P$ and $T \cap Q$.

7.5. Find an example in E^2 to show that the "Kirchberger number" $2n + 2 = 6$ in Exercise 7.4 cannot be reduced.

7.6. When two compact sets can be *strictly* separated by a hyperplane, then in fact there exist infinitely many hyperplanes that strictly separate them. When two compact sets can only be *weakly* separated by a hyperplane, then this hyperplane may be unique. Therefore, it is natural to ask what number $f(n)$ will make the following statement true:

Let P and Q be nonempty compact subsets of E^n. Suppose there exists a unique hyperplane separating P and Q. Then for some subset T of $f(n)$ or fewer points of $P \cup Q$ there exists a unique hyperplane separating $T \cap P$ and $T \cap Q$.

Find an example to show that no such number $f(n)$ exists.

7.7. (a) Use Kirchberger's Theorem 7.1 and the strict separation Theorem 4.13 to prove Caratheodory's Theorem 2.23 for compact sets.

(b) Use part (a) and Theorem 2.22 to prove Caratheodory's Theorem 2.23 for any subset S of E^n.

*7.8. Let P and Q be disjoint finite sets in E^n, and let x be a fixed point in $P \cup Q$. Prove that P and Q can be strictly separated by a hyperplane if for each set T consisting of the point x and $n + 1$ other points of $P \cup Q$, there exists a hyperplane that strictly separates $T \cap P$ and $T \cap Q$.

4

KIRCHBERGER-TYPE THEOREMS

In this chapter we consider several variations of Kirchberger's Theorem. There is a particular formulation of his result that motivates these variations:

Suppose P and Q are nonempty compact subsets of E^n. Then there exists a closed half-space S such that $S \supset P$ and $S \cap Q = \varnothing$ if and only if for each set T consisting of $n + 2$ or fewer points from $P \cup Q$ there exists a closed half-space S_T such that $S_T \supset (T \cap P)$ and $S_T \cap (T \cap Q) = \varnothing$.

It is easy to see that this formulation is equivalent to Kirchberger's Theorem 7.1. The following question immediately arises: If we replace S by some figure other than a closed half-space, what simple condition will ensure that we can find such a figure containing P and disjoint from Q?

In Section 8 we look at the case when S is a sphere. In Section 9 we consider generalized cylinders, and in Section 10 we investigate parallelotopes.

SECTION 8. SEPARATION BY A SPHERICAL SURFACE

The problem of replacing the "separating hyperplanes" in Kirchberger's Theorem by "separating spherical surfaces" was first posed by F. A. Valentine in 1964. At that time the critical number for E^2 was known to be five. The theorem which follows generalizes this result to E^n.

8.1. Definition. The spherical surface $S \equiv \{x: \|x - p\| = \alpha\}$ **strictly separates** the sets A and B if for all $a \in A$ and $b \in B$ one of the following holds:
 1. $\|a - p\| < \alpha$ and $\|b - p\| > \alpha$.
 2. $\|b - p\| < \alpha$ and $\|a - p\| > \alpha$.

8.2. Theorem. Let P and Q be nonempty compact subsets of E^n. Then there exists a spherical surface that strictly separates P and Q if and only if for each subset T of $n + 3$ or fewer points of $P \cup Q$ there exists a spherical surface that strictly separates $T \cap P$ and $T \cap Q$.

PROOF. Embed E^n in E^{n+1} and let Ω be a unit sphere in E^{n+1} which is tangent to E^n at an arbitrary point p. (See Figure 8.1.) Let π be the stereographic projection of E^n onto Ω based at the point antipodal to p in Ω. Suppose that given any subset T of $n + 3$ or fewer points of $P \cup Q$ there exists a spherical surface S in E^n which strictly separates $T \cap P$ and $T \cap Q$. That is, given any subset $\pi(T)$ of $n + 3$ or fewer points of $\pi(P \cup Q)$, there exists a spherical surface $\pi(S)$ on Ω that strictly separates $\pi(T \cap P)$ and $\pi(T \cap Q)$ on Ω. Now $\pi(S)$ is the intersection of some n-dimensional flat H with Ω, and H strictly separates $\pi(T \cap P)$ and $\pi(T \cap Q)$ in the $(n + 1)$-dimensional space E^{n+1}. Thus, by Kirchberger's Theorem 7.1, there exists an n-dimensional flat H_0 in E^{n+1} which strictly separates $\pi(P)$ and $\pi(Q)$. Furthermore, since any hyperplane parallel to H_0 and sufficiently close to it also strictly separates $\pi(P)$ and $\pi(Q)$, we may choose H_0 so that it does not contain the point antipodal to p in Ω. It follows that $\pi^{-1}(H_0 \cap \Omega)$ is a spherical surface in E^n that strictly separates P and Q.

The converse is immediate. ∎

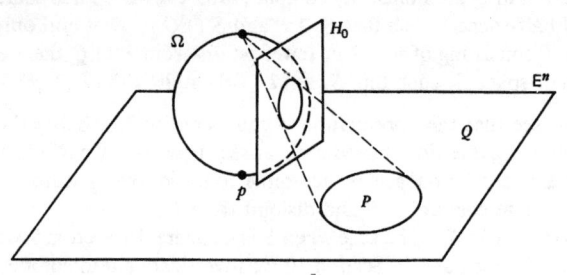

Figure 8.1.

The following example shows that Theorem 8.2 is a "best" theorem in the sense that the number $n + 3$ cannot be reduced.

8.3. Example. Let $P \equiv \{p_1, \ldots, p_{n+1}\}$ be the vertices of an n-dimensional simplex in E^n. Let Q consist of two points q_1 and q_2, where q_1 is in the relative interior of some face of the simplex conv P, and q_2 is outside the simplex and collinear with q_1 and the vertex opposite the face containing q_1. (See Figure 8.2 for $n = 2$.) If either point of Q is missing, then one can find a spherical surface S_1 around the remaining point that strictly separates it from P. If the point of P that lies between q_1 and q_2 is missing, then one can find a spherical surface S_2 around Q that strictly separates it from the remaining part of P. If one of the other points of P is missing, then one can find a spherical surface S_3 around the remaining part of P that strictly separates it from Q. Thus given any subset T of $n + 2$ or fewer points of $P \cup Q$, there exists a spherical surface

EXERCISES

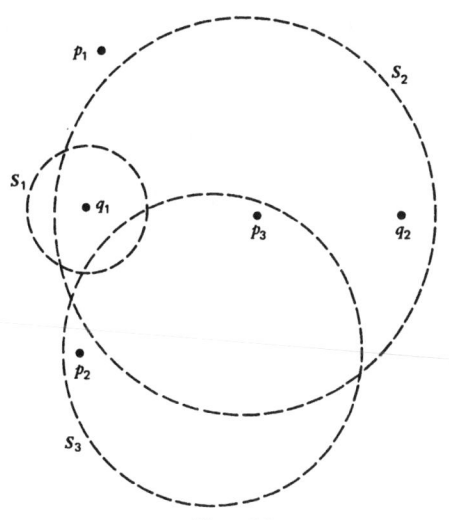

Figure 8.2.

that strictly separates $T \cap P$ and $T \cap Q$. But there does not exist a spherical surface that strictly separates P and Q since $P \cap \text{conv}\, Q \neq \emptyset$ and $Q \cap \text{conv}\, P \neq \emptyset$.

EXERCISES

8.1. Prove the following theorem for strict separation by a closed convex set: Let P and Q be nonempty compact subsets of E^n. Then there exists a closed convex set containing P and disjoint from Q iff for each set T consisting of $n + 2$ or fewer points from $P \cup Q$ there exists a closed convex set containing $T \cap P$ and disjoint from $T \cap Q$.

8.2. Find an example to show that the "Kirchberger number" $n + 2$ in Exercise 8.1 cannot be reduced.

8.3. Prove the following theorem for (weak) separation where the definition of a spherical surface is extended to include the limiting case of a hyperplane: Let P and Q be nonempty compact subsets of E^n. Then there exists a spherical surface that separates P and Q iff for each subset T of $2n + 4$ or fewer points of $P \cup Q$ there exists a spherical surface that separates $T \cap P$ and $T \cap Q$.

8.4. Find an example to show that the "Kirchberger number" $2n + 4$ in Exercise 8.3 cannot be reduced.

*8.5. Prove the following theorem of Dines and McCoy (1933): "Let A be a subset of E^n. If $y \in \text{int conv}\, A$, then there exist points x_i in A ($i = 1, \ldots, m \leq 2n$) such that $y \in \text{int conv}\{x_1, \ldots, x_m\}$."

8.6. Use Exercise 8.5 to prove the following theorem for (weak) separation by a closed convex set: Let P and Q be nonempty subsets of E^n. Then there exists a closed convex set S such that $P \subset S$ and $Q \cap \text{int } S = \emptyset$ iff for each subset T of $2n + 1$ or fewer points of $P \cup Q$ there exists a closed convex set S_T such that $(P \cap T) \subset S_T$ and $(Q \cap T) \cap \text{int } S_T = \emptyset$.

8.7. Find an example to show that the "Kirchberger number" $2n + 1$ in Exercise 8.6 cannot be reduced.

SECTION 9. SEPARATION BY A CYLINDER

The question of separating two compact sets by a "cylinder" was raised at the end of Section 4. We return to this topic now and begin by stating precisely what we mean by a cylinder.

9.1. Definition. Let A be a nonempty compact convex subset of E^n and let F be a k-dimensional subspace ($0 \le k \le n$). Then $C \equiv A + F \equiv \{a + f: a \in A \text{ and } f \in F\}$ is called the ***k*-cylinder generated by A and F**.

Notice that the 0-cylinder generated by A is precisely A. The 1-cylinder generated by A and a line through the origin is just a cylinder in the sense of the usual definition. An $(n - 1)$-cylinder is a pair of parallel hyperplanes and the region between them, that is, a slab. The n-cylinder generated by A is precisely E^n. The case where A is a closed ball in E^3 is illustrated in Figure 9.1.

In light of the earlier theorems on strictly separating two compact sets by a hyperplane and by a sphere, it is natural to ask what number $f(n, k)$ will make the following statement true:

> Let P and Q be nonempty compact subsets of E^n. Then there exists a k-cylinder of the form $C \equiv (\text{conv } P) + F$ such that $C \cap Q = \emptyset$ if and only if for each subset T of $f(n, k)$ or fewer points of $P \cup Q$ there exists a k-cylinder of the form $C_T \equiv [\text{conv}(T \cap P)] + F_T$ such that $C_T \cap (T \cap Q) = \emptyset$.

It is perhaps surprising to discover that in general no such number exists. Indeed, we can even find an example to show this in the plane E^2.

9.2. Example. Let m be any even positive integer. We shall construct a finite set Q and a compact set P in the plane E^2 that satisfy the following:

(a) Every 1-cylinder of the form $C \equiv (\text{conv } P) + L$, where L is a line through the origin, has a nonempty intersection with Q.

(b) Given any subset T of m points of Q, there exists a 1-cylinder $C_T \equiv (\text{conv } P) + L_T$ such that $C_T \cap T = \emptyset$.

We begin by noting that condition (b) is even stronger than we need, since T is a subset of Q and C_T contains all of P.

SEPARATION BY A CYLINDER

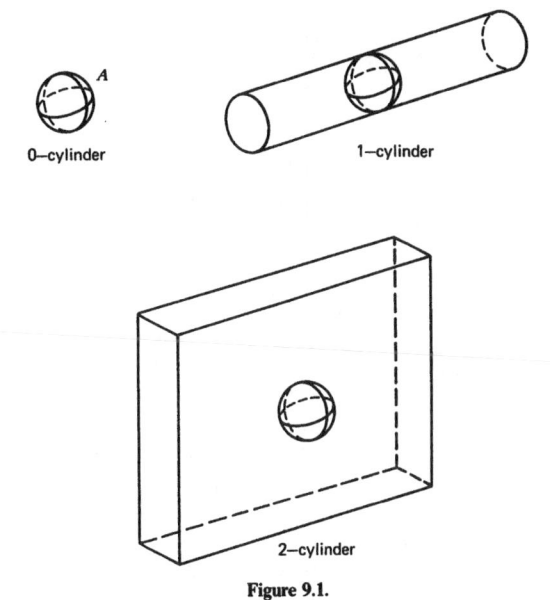

Figure 9.1.

Let P be the unit circle about the origin, and let $d_1, \ldots, d_{m+1}$ be a set of $m+1$ rays emanating from the origin which divide P into $m+1$ equal parts. The set $Q \equiv \{q_1, \ldots, q_{m+1}\}$ will be of the form $q_i \equiv S_r \cap d_i$ ($i = 1, \ldots, m+1$), where S_r is the circle of radius r about the origin. We now choose r so that conditions (a) and (b) are satisfied.

If we delete any one element of Q, say q_j, then in order to find a 1-cylinder generated by P which will miss Q, it is sufficient that the distance between any two members of Q be greater than 2. For then, the 1-cylinder $(\operatorname{conv} P) + L_j$ misses Q, where L_j is the line through the origin containing d_j. A simple computation shows that this can be accomplished if and only if

$$r > \csc\left(\frac{180}{m+1}\right)^\circ.$$

On the other hand, if it were possible to find a 1-cylinder containing P and missing all of Q, then $m/2 + 1$ points of Q would have to lie on one side of the cylinder. A simple computation shows that this can be accomplished if and only if

$$r < \csc\left(\frac{90}{m+1}\right)^\circ.$$

Now $\csc[180/(m+1)]^\circ$ is strictly less than $\csc([90/(m+1)]^\circ$ for $m > 1$,

KIRCHBERGER-TYPE THEOREMS

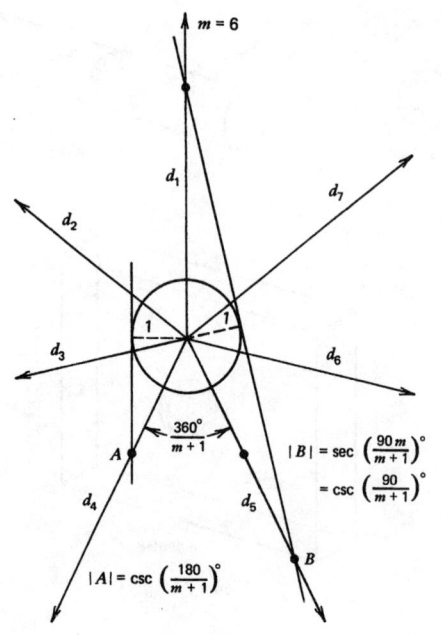

Figure 9.2.

and so by choosing r so that

$$\csc\left(\frac{180}{m+1}\right)^\circ < r < \csc\left(\frac{90}{m+1}\right)^\circ$$

and by defining $Q \equiv \{S_r \cap d_i: i = 1,\ldots, m+1\}$, the sets P and Q satisfy both conditions (a) and (b). The case when $m = 6$ is illustrated in Figure 9.2.

The preceding example, however, does not mean that a similar theorem for strictly separating two compact sets by a k-cylinder is nonexistent. The critical change that yields the appropriate theorem involves strictly separating the subsets with hyperplanes instead of k-cylinders. This produces Theorem 9.5 after a preliminary definition and a lemma.

9.3. Definition. A subset K of the surface of an n-dimensional ball is said to be **strongly convex** if K does not contain antipodal points and does contain, with each pair of its points, the small arc of the great circle determined by them.

9.4. Lemma. Let Ω be the surface of the unit ball centered at the origin in E^n, and let $\mathcal{F} \equiv \{A_\alpha: \alpha \in \mathcal{C}\}$ be a family of compact strongly convex subsets

SEPARATION BY A CYLINDER

of Ω. If each n or fewer members of $\mathcal{F}$ have a point in common, then there exists a pair of antipodal points $\{y, -y\}$ in Ω such that each A_α ($\alpha \in \mathcal{C}$) intersects $\{y, -y\}$.

PROOF. Since each A_α ($\alpha \in \mathcal{C}$) is a compact strongly convex subset of Ω, it follows that each $\text{conv} A_\alpha$ ($\alpha \in \mathcal{C}$) is a compact convex subset of E^n which does not contain the origin. If every n or fewer members of $\mathcal{F}$ have a point in common, then Theorem 6.8 applied to the family $\{\text{conv} A_\alpha : \alpha \in \mathcal{C}\}$ implies that there exists a line L through the origin which intersects each $\text{conv} A_\alpha$ ($\alpha \in \mathcal{C}$). But then L also intersects each member of $\mathcal{F}$, and letting $\{y, -y\} = L \cap \Omega$, we have the desired pair of antipodal points. ∎

9.5. Theorem. Let P and Q be nonempty compact subsets of E^n. Suppose for a fixed integer k ($1 \leq k \leq n$) that each subset of k or fewer points of Q can be strictly separated from P by a hyperplane. Then given any k-cylinder of the form $C \equiv (\text{conv} P) + F$ there exists a $(k-1)$-cylinder of the form $D \equiv (\text{conv} P) + F_1$ such that $D \subset C$ and $D \cap Q = \emptyset$.

PROOF. Let $\delta = \inf\{d(\text{conv} T, \text{conv} P) : T \text{ is a subset of } k \text{ or fewer points of } Q\}$. Since P and Q are compact, it follows that $\delta > 0$. Given a k-cylinder $C \equiv (\text{conv} P) + F$, if $Q \cap C = \emptyset$, then the result follows directly. If $Q \cap C \neq \emptyset$, then we let Ω be the intersection of F and the surface of the unit ball about the origin in E^n. For each point $w \in \Omega$ we define r_w to be the ray from the origin through w, and F_w to be the $(k-1)$-dimensional subspace contained in F which is perpendicular to r_w. Then for each point $q \in Q \cap C$, we define

$$A_q \equiv \{w \in \Omega : S_q \text{ is contained in the component of } C \sim [(\text{conv} P) + F_w] \text{ that intersects } (\text{conv} P) + r_w\}$$

where $S_q \equiv \{x : \|x - q\| < \delta/2\}$. (See Figure 9.3.)

First, we claim that for each $q \in Q \cap C$, A_q is a compact strongly convex subset of Ω. To see this we pick an A_q and define, for each $w \in A_q$, $\mathcal{G}_w$ to be the component of $C \sim [(\text{conv} P) + F_w]$ which intersects $(\text{conv} P) + r_w$. Thus given two distinct points w and w' in A_q, we must show that $S_q \subset \mathcal{G}_x$ where x is a point on the small arc of the great circle determined by w and w'. Now $S_q \subset \mathcal{G}_w \cap \mathcal{G}_{w'}$ and $\mathcal{G}_w \cap \mathcal{G}_{w'}$ is contained in $\mathcal{G}_x$, so $S_q \subset \mathcal{G}_x$. Furthermore, since S_q is open, A_q is a compact subset of some open hemisphere and thus contains no antipodal points.

Second, we claim that if $q_1, \ldots, q_m$ ($1 \leq m \leq k$) are any m points in $Q \cap C$, then $\bigcap_{i=1}^{m} A_{q_i} \neq \emptyset$. To see this we note that $d(\text{conv}\{q_1, \ldots, q_m\}, \text{conv} p) \geq \delta$. Thus $d(\text{conv}\{S_{q_1}, \ldots, S_{q_m}\}, \text{conv} P) > \delta/2$ and so there exists a hyperplane H strictly separating $\{S_{q_1}, \ldots, S_{q_m}\}$ and P. Let H' be the $(n-1)$-dimensional subspace parallel to H, and let $G \equiv H' \cap F$. Now $F \not\subset H'$ since H separates P and $\{S_{q_1}, \ldots, S_{q_m}\}$, and so G is a $(k-1)$-dimensional subspace. It follows that $\{S_{q_1}, \ldots, S_{q_m}\}$ is contained in one of the two components of $C \sim [(\text{conv} P) +

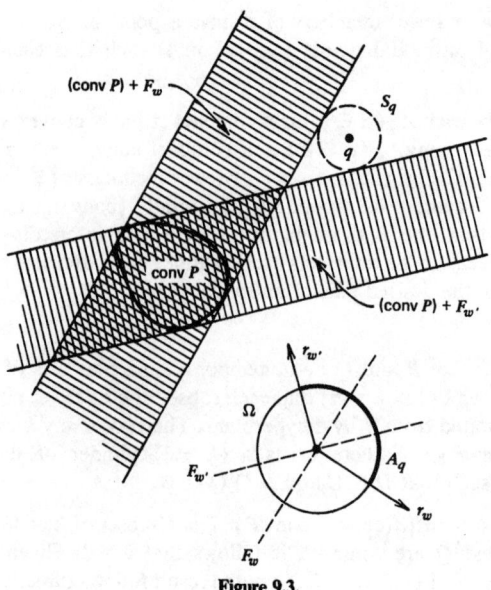

Figure 9.3.

G], and we may choose $w \in \Omega$ so that r_w is perpendicular to G and $(\operatorname{conv} P) + r_w$ intersects the component that contains $\{S_{q_1}, \ldots, S_{q_m}\}$. Then $w \in \bigcap_{i=1}^{m} A_{q_i}$.

Combining the two claims and Lemma 9.4, we see that there exists a pair of antipodal points $\{y, -y\}$ in Ω such that each A_q ($q \in Q \cap C$) intersects $\{y, -y\}$. It follows that the $(k-1)$-cylinder $(\operatorname{conv} P) + F_y$ has an empty intersection with Q. ∎

We can now answer the question raised at the end of Section 4. Specifically, for compact subsets P and Q of E^n we have the following: If every k points of Q (where $1 \leq k \leq n$) can be strictly separated from P by a hyperplane, then there exists a $(k-1)$-cylinder containing P which is disjoint from Q.

Consider, for example, the situation in E^3. If each single point of Q can be strictly separated from P by a hyperplane, then Q is outside the convex hull of P. (See Figure 9.4a.) If every two points of Q can be strictly separated from P by a hyperplane, then there exists a 1-cylinder with P on the inside and Q on the outside. (See Figure 9.4b.) If every three points of Q can be strictly separated from P by a hyperplane, then there exist two parallel hyperplanes with P between them and Q on the outside. (See Figure 9.4c.) If every four points of Q can be strictly separated from P by a hyperplane, then (by Caratheodory's Theorem) $\operatorname{conv} P \cap \operatorname{conv} Q = \varnothing$, so that P and Q can be strictly separated by a hyperplane. (See Figure 9.4d.)

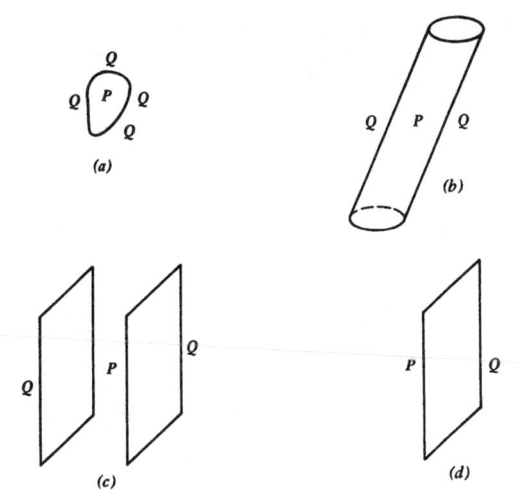

Figure 9.4.

EXERCISES

9.1. Find examples to show that the hypotheses of Theorem 9.5 cannot be weakened and still yield the same conclusion. Specifically, find the following in E^2:
   (a) Two compact sets P and Q with the property that each point of Q can be strictly separated from P by a hyperplane but such that there exists no 1-cylinder containing P which misses Q.
   (b) A compact set P and a closed set Q with the property that each two points of Q can be strictly separated from P by a hyperplane but such that there exists no 1-cylinder containing P which misses Q.
   (c) A compact set P and a bounded set Q with the property that each two points of Q can be strictly separated from P by a hyperplane but such that there exists no 1-cylinder containing P which misses Q.

9.2. Use Theorem 9.5 to prove the following theorem of Hanner and Rådström (1951): "Let Q be a compact subset of E^n and suppose p is a point of E^n which does not have the n-point simplicial property with respect to Q. Then there exists a hyperplane through p which misses Q."

9.3. Let A be a nonempty compact convex subset of E^n and let $C \equiv A + F$ be any fixed k-cylinder generated by A ($1 \leq k \leq n$). Prove that dim $A = n$ iff for each $(k-1)$-cylinder of the form $D \equiv A + F_1$ with $D \subset C$, it is true that dim $D = n$.

9.4. Let P and Q be nonempty compact subsets of E^n. Suppose for a fixed integer k ($1 \leq k \leq n$) that each subset of k or fewer points of Q can be (weakly) separated from P by a hyperplane. Prove that given any k-cylinder of the form $C \equiv (\operatorname{conv} P) + F$, there exists a $(k-1)$-cylinder of the form $D \equiv (\operatorname{conv} P) + F_1$ such that $D \subset C$ and $Q \cap \operatorname{int} D = \varnothing$.

9.5. Let P and Q be nonempty compact subsets of E^n. Suppose for a fixed integer k ($2 \leq k \leq n+1$) that each subset of k or fewer points of Q can be strictly separated from P by a hyperplane. Prove that given any $(k-1)$-cylinder of the form $D \equiv (\operatorname{conv} P) + F_1$, there exists a $(k-2)$-cylinder of the form $E \equiv (\operatorname{conv} P) + F_2$ such that $E \subset D$ and $Q \cap D$ is contained in one of the two connected components of $D \sim E$.

9.6. Use Exercise 9.5 to prove Kirchberger's Theorem 7.1.

†9.7. Let P be a sphere in E^3 and let Q be a nonempty compact subset of E^3. Suppose that each subset of two or fewer points of Q can be strictly separated from P by a hyperplane. What additional hypothesis will guarantee that there exists a 2-cylinder containing P and disjoint from Q?

9.8. Let Ω be the surface of an n-dimensional ball in E^n, and let $\mathcal{F}$ be a family of compact strongly convex subsets of Ω. Suppose that no $n+1$ of the members of $\mathcal{F}$ cover Ω and that the intersection of any n of them is nonempty. Prove that the intersection of all the members of $\mathcal{F}$ is nonempty.

9.9. Let Ω be the surface of an n-dimensional ball in E^n, and let $\mathcal{F}$ be a family of $n+2$ or more strongly convex subsets of Ω. Suppose that the intersection of any $n+1$ of the members of $\mathcal{F}$ is empty. Prove that the intersection of some collection of n of the members of $\mathcal{F}$ is empty.

SECTION 10. SEPARATION BY A PARALLELOTOPE

Chapter 4 concludes with an investigation of the question of separating two planar sets by a parallelogram, two spatial sets by a parallelepiped, and in general, two n-dimensional sets by a parallelotope (see Definition 21.9). This type of separation is useful because the separating set can be described by a system of simultaneous linear inequalities. It also enables us to derive a special modification of Caratheodory's Theorem. We begin with a definition.

10.1. Definition. Given a basis $\beta \equiv \{b_1, b_2, \ldots, b_n\}$ for E^n, let H_i ($i = 1, \ldots, n$) be the $(n-1)$-dimensional subspace spanned by $\beta \sim \{b_i\}$. A β-box is a (possibly degenerate) compact solid parallelotope in which each side is parallel to one of the H_i ($i = 1, \ldots, n$).

Algebraically, for any point $x \equiv \sum_{i=1}^n \alpha_i b_i$ we define the ith coordinate function π_i by $\pi_i(x) = \alpha_i$. Then the null space of π_i is H_i and a β-box is the set

SEPARATION BY A PARALLELOTOPE

of all points x that satisfy a system of linear inequalities of the form

$$m_1 \leq \pi_1(x) \leq M_1$$
$$m_2 \leq \pi_2(x) \leq M_2 \quad \text{for real numbers } m_i \leq M_i \, (i = 1, \ldots, n).$$
$$\vdots$$
$$m_n \leq \pi_n(x) \leq M_n$$

Given any nonempty compact subset P of E^n, there is a unique minimal β-box containing P. Specifically, take

$$m_i = \inf_{x \in P} \pi_i(x) \quad \text{and} \quad M_i = \sup_{x \in P} \pi_i(x) \quad (i = 1, \ldots, n).$$

The β-box so determined clearly contains P and is contained in any other β-box containing P. An example in E^2 is illustrated in Figure 10.1.

10.2. Theorem. Let P and Q be nonempty compact subsets of E^n ($n \geq 2$). Then for any basis β of E^n, the following three conditions are equivalent:
  (a) There exists a β-box B such that $P \subset B$ and $Q \cap B = \varnothing$.
  (b) For each subset S of $n+1$ or fewer points of $P \cup Q$, there exists a β-box B_S such that $(P \cap S) \subset B_S$ and $(Q \cap S) \cap B_S = \varnothing$.
  (c) For each subset T of n or fewer points of P, the minimal β-box containing T is disjoint from Q.

PROOF. The implication (a) $\Rightarrow$ (b) is obvious. To see that (b) implies (c), suppose that (c) does not hold. Then there exists a subset T of n or fewer points of P such that Q intersects the minimal β-box B_T containing T. Let $q \in Q \cap B_T$. Then $S \equiv \{q\} \cup T$ is a subset of $n+1$ or fewer points of $P \cup Q$ such that there is no β-box containing $P \cap S = T$ and disjoint from $Q \cap S = \{q\}$. Thus condition (b) does not hold and we have proved the contrapositive of (b) $\Rightarrow$ (c).

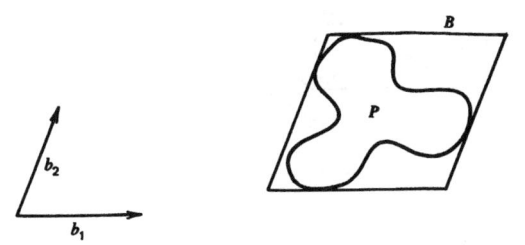

Figure 10.1.

The proof that (c) implies (a) is by an induction on the dimension n of the space. Suppose that $n > 2$ and that the theorem holds for spaces of dimension m, where $2 \leqslant m < n$. Suppose further that condition (a) does not hold. Then the minimal β-box B containing P is not disjoint from Q. That is, there exists a point q in $Q \cap B$. We must show that condition (c) does not hold.

Define a projection mapping f from E^n onto E^{n-1} by

$$f(\alpha_1 b_1 + \alpha_2 b_2 + \cdots + \alpha_n b_n) \equiv \alpha_2 b_2 + \cdots + \alpha_n b_n.$$

Then $\beta' \equiv \{b_2, \ldots, b_n\}$ is a basis for E^{n-1} and $f(B)$ is the minimal β'-box containing $f(P)$. Now $f(q) \in f(B)$, so the theorem applied to the compact sets $f(P)$ and $f(q)$ in E^{n-1} implies that there exists a subset T' of $n - 1$ or fewer points of $f(P)$ such that $f(q)$ is in the minimal β'-box $B_{T'}$ containing T'. Let T be a subset of $n - 1$ or fewer points of P such that $f(T) = T'$. Since $f(q) \in B_{T'}$ and $\pi_i \circ f = \pi_i$ for $i = 2, \ldots, n$, it follows that

$$\inf_{x \in T} \pi_i(x) \leqslant \pi_i(q) \leqslant \sup_{x \in T} \pi_i(x)$$

for $i = 2, \ldots, n$. If these inequalities also hold for $i = 1$, then q is in the minimal β-box B_T containing T, and we are done. If they do not hold for $i = 1$, then either

$$\pi_1(q) > \sup_{x \in T} \pi_1(x) \quad \text{or} \quad \pi_1(q) < \inf_{x \in T} \pi_1(x).$$

If $\pi_1(q) > \sup_{x \in T} \pi_1(x)$, then since P is compact we may choose a point p in P such that $\pi_1(p) = \sup_{x \in P} \pi_1(x)$. Since $q \in B$ we have $\pi_1(q) \leqslant \sup_{x \in P} \pi_1(x)$ and

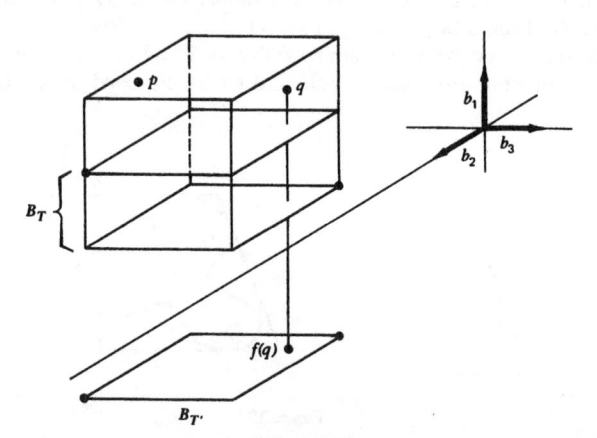

Figure 10.2.

SEPARATION BY A PARALLELOTOPE

it follows that

$$\inf_{x \in T \cup \{p\}} \pi_1(x) \leq \pi_1(q) \leq \sup_{x \in T \cup \{p\}} \pi_1(x).$$

(See Figure 10.2.) Similarly, if $\pi_1(q) < \inf_{x \in T} \pi_1(x)$, we choose a point p in P such that $\pi_1(p) = \inf_{x \in P} \pi_1(x)$ and obtain the same result. Thus in both cases we have found a subset of n or fewer points of P whose minimal β-box contains the point q. Therefore, condition (c) does not hold and we have established the induction part of the proof.

Finally we must show that (c) implies (a) when $n = 2$. Assume that condition (c) holds and let B be the minimal β-box containing P. Since B is minimal, we may select a point of P on each of the four sides of the parallelogram B. Call these points p_1, p_2, p_3, and p_4. (They are not necessarily unique or distinct.) Suppose that there exists a point q in $Q \cap B$. We must find two points of P such that q lies in the minimal β-box containing those two points. Now for each pair of points p_i and p_j ($i \neq j$), let B_{ij} be the minimal β-box containing $\{p_i, p_j\}$. Then it is easy to see that $B = \cup_{i \neq j} B_{ij}$. (In fact, B can be expressed as the union of five or fewer of the B_{ij}'s. For example, see Figure 10.3.) Thus $q \in B_{ij}$ for some particular i and j. This contradicts our assumption that condition (c) holds and so Q and B must be disjoint. ∎

If we rephrase Theorem 10.2 in terms of systems of inequalities we obtain the following corollary. (See the comments prior to Theorem 10.2 for the notation.)

10.3. Corollary. Let P be a nonempty compact subset of E^n and suppose that a point q in E^n satisfies the following system of inequalities:

$$\inf_{x \in P} \pi_i(x) \leq \pi_i(q) \leq \sup_{x \in P} \pi_i(x) \quad \text{for } i = 1, \ldots, n.$$

Then there exists a subset T of n or fewer points of P such that

$$\inf_{x \in T} \pi_i(x) \leq \pi_i(q) \leq \sup_{x \in T} \pi_i(x) \quad \text{for } i = 1, \ldots, n.$$

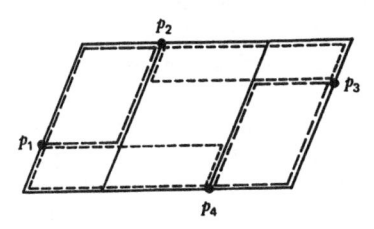

$B = B_{12} \cup B_{14} \cup B_{23} \cup B_{24} \cup B_{34}$ **Figure 10.3.**

It is not surprising in Corollary 10.3 that there exists a finite subset T with the desired property. Indeed we can choose a maximizing point and a minimizing point in P for each coordinate and readily obtain a subset T of $2n$ points in P that work. The force of the corollary is that there exist n points in P with the same property.

There is a special case included in Theorem 10.2 that is of particular interest because of its similarity to Caratheodory's Theorem 2.23. In light of the separation theorem, Caratheodory's Theorem may be rephrased as:

Let P be a compact subset of E^n and let q be a point in E^n. Suppose that given any subset T of $n + 1$ or fewer points of P there exists a hyperplane which strictly separates T and $\{q\}$. Then q is not in the convex hull of P.

The following corollary shows that if the orientation of the separating hyperplanes is properly restricted, then it suffices to consider subsets of n or fewer points of P.

10.4. Corollary. Let P be a nonempty compact subset of E^n ($n \geq 2$) and let q be a point in E^n. Given a basis $\beta \equiv \{b_1, \ldots, b_n\}$ of E^n, let H_i ($i = 1, \ldots, n$) be the $(n - 1)$-dimensional subspace spanned by $\beta \sim \{b_i\}$. Suppose that given any subset T of n or fewer points of P there exists a hyperplane that is parallel to one of the H_i ($i = 1, \ldots, n$) and that strictly separates T and $\{q\}$. Then q is not in the convex hull of P.

PROOF. The hypotheses imply that for each subset T of n or fewer points of P there exists a β-box containing T and disjoint from $\{q\}$. Applying Theorem 10.2, we conclude that there exists a β-box containing P and disjoint from $\{q\}$, and so $q \notin \operatorname{conv} P$. ∎

EXERCISES

10.1. Prove the following theorem: Let H be a fixed hyperplane in E^n and suppose P and Q are compact subsets of E^n. If for each subset T of three or fewer points of $P \cup Q$ there exists a hyperplane parallel to H which strictly separates $T \cap P$ and $T \cap Q$, then there exists a hyperplane parallel to H which strictly separates P and Q.

10.2. Find an example in E^2 to show that the number n in Corollary 10.4 cannot be reduced to $n - 1$.

10.3. In the proof of Theorem 10.2, it is crucial that all of the parallelotopes are similarly oriented. What happens if this restriction is dropped? Show that under these conditions there is no Kirchberger-type number. Specifically, show that given any positive integer k, there exist nonempty compact subsets P and Q of E^2 with the following property: Given any subset T of k or fewer points of P, there exists a parallelogram containing T which is disjoint from Q. But there is no parallelogram containing all of P which is disjoint from Q.

†**10.4.** Let P and Q be nonempty compact subsets of E^n. What conditions on P and Q will guarantee the existence of a simplex (or a cone, etc.) containing P and disjoint from Q?

5

SPECIAL TOPICS IN E^2

Up to this point we have been studying convex sets in the full generality of n-dimensional Euclidean space. We now turn our attention to some special topics that are more appropriately considered in the setting of the Euclidean plane. Although most of these results have been generalized to E^3 (and many of them to E^n), their historical significance has been primarily in E^2.

SECTION 11. SETS OF CONSTANT WIDTH

Many interesting physical applications of convex sets are related to the concept of width and sets whose width is the same in all directions.

11.1. Definition. Let S be a nonempty compact subset of E^2 and let ℓ be a line (or line segment) in E^2. The **width of S in the direction of ℓ** is the distance between the two parallel lines of support to S that are perpendicular to ℓ and that contain S between them. (See Figure 11.1.)

11.2. Definition. Let S be a nonempty bounded subset of E^2. The **diameter** d of S is defined to be the number

$$d \equiv \sup_{\substack{x \in S \\ y \in S}} \|x - y\|.$$

The relationship between the diameter of a set and its width is established in the following theorem.

11.3. Theorem. The diameter of a nonempty compact set S in E^2 is equal to the maximum width of S.

PROOF. Let m be the maximum width of S and let H_1 and H_2 be parallel supporting lines of S at a distance m apart. If x and y are points of S lying on H_1 and H_2, respectively, we claim that $\|x - y\| = m$. If $\overline{xy}$ is perpendicular to H_1, then certainly $\|x - y\| = m$. Thus suppose $\overline{xy}$ is not perpendicular to H_1,

SETS OF CONSTANT WIDTH

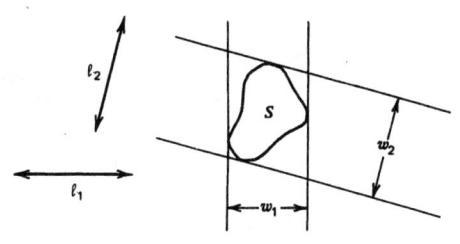

Figure 11.1.

and let H_3 and H_4 be parallel supporting lines of S that are perpendicular to $\overline{xy}$. (See Figure 11.2.) If m' is the width of S in the direction of $\overline{xy}$, then we have

$$m < \|x - y\| \leq m'.$$

This contradicts the maximality of m and shows that $\|x - y\| = m$.

Finally, let p and q be any two points of S and let w be the width of S in the direction of $\overline{pq}$. Then we have

$$\|x - y\| = m \geq w \geq \|p - q\|,$$

so that m is the diameter of S. ∎

11.4. Corollary. Let S be a nonempty compact subset of E^2 and let H_1 and H_2 be parallel supporting lines of S in a direction of maximum width.
  (a) If x is any point of $H_1 \cap S$, then the line ℓ through x perpendicular to H_1 intersects H_2 in a point y which belongs to S.
  (b) $H_1 \cap S$ and $H_2 \cap S$ each contain exactly one point, and the line through these two points is perpendicular to H_1 and H_2.

PROOF. Exercise 11.4 ∎

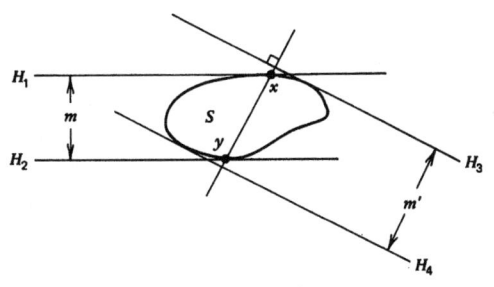

Figure 11.2.

In general, a set has different widths in different directions. If it happens that the width of a set is the same for all directions then the set is said to be of **constant width**. The simplest example of a set with this property is of course a circle. Noncircular sets of constant width were first studied by Euler in the 1770s, and since then by many other mathematicians. Among the early contributors was F. Reuleaux, for whom the following "Reuleaux triangle" is named:

Let $\triangle ABC$ be an equilateral triangle with the length of each side equal to w. With each vertex as center, draw the smaller arc of radius w joining the other two vertices. The union of these three arcs is the boundary of a Reuleaux triangle. Clearly it is a noncircular set of constant width w. Indeed, given any two parallel supporting lines, one of them will pass through a vertex of $\triangle ABC$, while the other will be tangent to the opposite circular arc. Thus the distance between them is w. (See Figure 11.3.)

11.5. Definition. Let S be a compact subset of E^2. A point x in the boundary of S is called a **corner point** (or **vertex**) if there exists more than one line of support to S at x.

While a Reuleaux triangle has corner points, other noncircular sets of constant width do not. For example, if R is a Reuleaux triangle of width w and U is a circle of radius h centered at the origin, then $R + U$ is a set of constant width $w + 2h$ which has no corners. Another way to construct $R + U$ is by extending each side of the triangle ABC a distance h in both directions. (See Figure 11.4.) With each vertex as center and with radius h draw the arcs $A'A''$, $B'B''$, and $C'C''$. Then with each vertex as center and with radius $w + h$, draw the arcs $A''B'$, $B''C'$, and $C''A'$. The six arcs form the boundary of $R + U$.

11.6. Theorem. Let S be a nonempty compact convex subset of E^2 that has constant width w. If $x \notin S$, then the diameter of $S \cup \{x\}$ is greater than w.

PROOF. Let y be the point of S closest to x and let H be the line through y perpendicular to $\overline{xy}$. (See Figure 11.5.) Then H is a supporting line of S and there exists a point z in bd S such that $y \in \text{relint } \overline{xz}$. It follows from Corollary

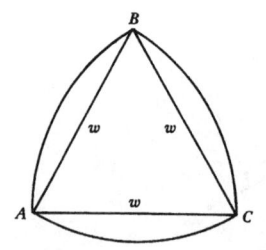

Figure 11.3.

SETS OF CONSTANT WIDTH

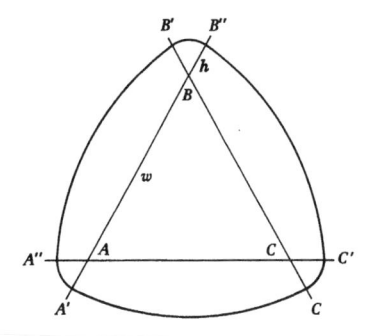

Figure 11.4.

11.4 that $\|y - z\| = w$. But then $\|x - z\| = \|x - y\| + \|y - z\| > w$, and the diameter of $S \cup \{x\}$ must be greater than w. ∎

The converse to Theorem 11.6 is also true, although we will not prove it here. (See Exercise 11.12.) Together they provide an interesting characterization of compact convex sets of constant width: they are precisely the sets that are complete in the sense that no point can be added to them without increasing their diameter.

11.7. Definition. A circle of largest diameter that lies entirely in a compact convex set S is called an **incircle** of S. The circle of smallest diameter that encloses S is called the **circumcircle**.

In general, a set may have many incircles, but the circumcircle is always unique. (See Exercise 11.6.) For sets of constant width, however, both the incircle and the circumcircle are unique—in fact, they are concentric. This is proved in the following theorem under the assumption that both circles exist.

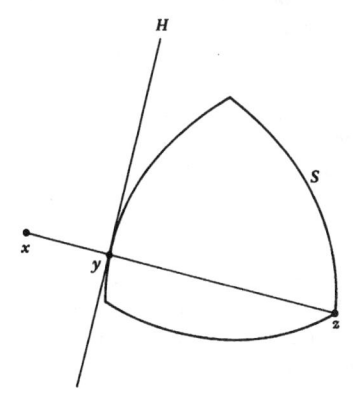

Figure 11.5.

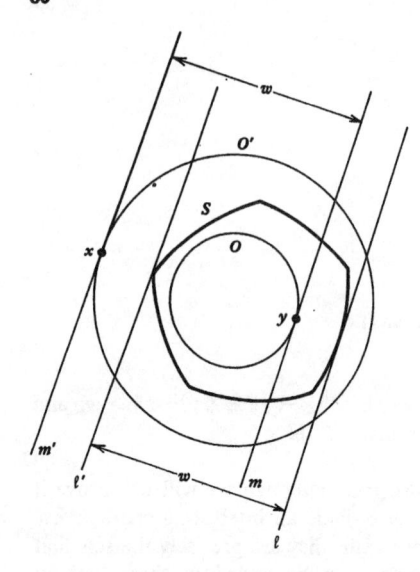

Figure 11.6.

The proof of the existence of an incircle and a circumcircle for compact sets will be given in Chapter 6.

11.8. Theorem. The incircle and the circumcircle of a compact convex set of constant width w are concentric, and the sum of their radii is equal to w.

PROOF. Let S be a compact convex set of constant width w and let O be a circle of radius r that is contained in S. Also, let O' be a circle of radius $w - r$ concentric with O. (See Figure 11.6.) We claim that O' contains S.

To see this let x be any point of O', m' the tangent line to O' at x, and m the tangent to O parallel to m' and at distance w from m'. Denote by y the point at which m is tangent to O. Also, let ℓ and ℓ' be the supporting lines of S parallel to m and m'. The distance between ℓ and ℓ' is w, the same as that between m and m'. Since y lies in S, the line m does not lie outside the strip formed by the lines ℓ and ℓ'. Thus m', at distance w from m, does not lie between ℓ and ℓ'. Hence the point x on m' is not in the interior of S, as we wished to show.

Similarly, if O' is a circle of radius R enclosing the set S and if O is a circle concentric with O' and with radius $w - R$, then we may conclude that O is contained in S.

Now let O_c be the circumcircle and O_i an incircle of S. Let R and r be their respective radii. Then R cannot be greater than $w - r$, since the circle of radius $w - r$ concentric with O_i encloses S and the circumcircle has the smallest radius of all circles enclosing S. On the other hand, R cannot be less than $w - r$, since otherwise a circle of radius $w - R$ concentric with O_c would lie inside S and would have a radius greater than r. Therefore, $R + r = w$.

SETS OF CONSTANT WIDTH

Finally, suppose that O_i and O_c are not concentric. Then the circle O' of radius R and concentric with O_i also encloses S. This contradicts the uniqueness of the circumcircle. ∎

The final two theorems in this section deal with the perimeter and the area of convex sets. For nonconvex sets it is difficult to define the concepts of length and area. But since compact convex sets may always be approximated by inscribed polygons (cf. Theorem 22.4), we have an effective means of defining their size. This is discussed more fully in the n-dimensional case in Section 22, but for now the following two definitions will suffice.

11.9. Definition. Let S be a compact convex subset of E^2 such that int $S \neq \emptyset$. The **perimeter** of S is the length of its boundary and is equal to the supremum of the perimeters of all the inscribed convex polygons.

11.10. Definition. Let S be a compact convex subset of E^2 such that int $S \neq \emptyset$. The **area** of S is equal to the supremum of the areas of all the inscribed convex polygons.

In the preceding definitions we could have used the infimum of circumscribed polygons about S instead of the supremum of inscribed polygons (see Theorem 22.5). In both cases, as the number of sides increases and the side length decreases, the area and the perimeter of the polygons approach that of the convex set S.

The simplest set of constant width w is a circle with diameter w. Its perimeter is of course equal to πw. It is not difficult to show (Exercise 11.1) that the Reuleaux triangle of width w also has perimeter πw. The following theorem of Barbier makes the remarkable assertion that every set of constant width w has perimeter of πw. The proofs of Barbier's Theorem and of Theorem 11.8 follow the approach used by Yaglom and Boltyanskii (1961). For another approach to Barbier's Theorem, see Kelly (1963).

11.11. Theorem (Barbier). Let S be any compact convex set in E^2 having constant width w. Then the perimeter of S is equal to πw.

PROOF. Let O be a circle of diameter w. We claim that any two equiangular polygons with 2^n sides circumscribed about O and S have the same perimeter. The proof is by induction on n. It is clear that two squares circumscribed about O and S have the same perimeter; in fact, these squares are congruent.

Suppose it is known that all equiangular polygons of 2^n sides circumscribed about O and S have equal perimeters. Consider two adjacent sides, say $\overline{be}$ and $\overline{bf}$, of an equiangular polygon of 2^n sides and the two opposite sides $\overline{dg}$ and $\overline{dh}$. (See Figure 11.7.) The extensions of these sides form a parallelogram $abcd$ with equal altitudes—that is, a rhombus. We carry out this construction for both O and S. The resulting rhombuses $abcd$ (circumscribed about O) and $a'b'c'd'$

82 SPECIAL TOPICS IN E²

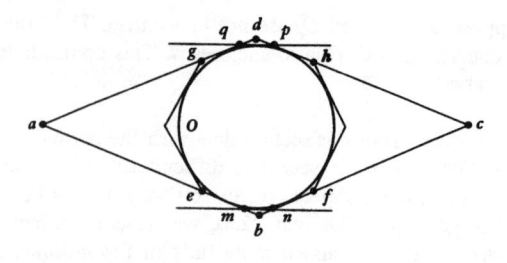

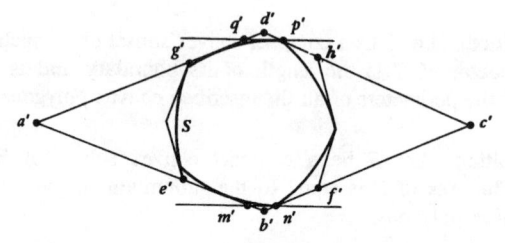

Figure 11.7.

(circumscribed about S) are congruent since they have equal angles and equal altitudes. We also draw supporting lines $\overline{mn}$ and $\overline{pq}$ to the circle O perpendicular to the diagonal $\overline{bd}$, and supporting lines $\overline{m'n'}$ and $\overline{p'q'}$ to the set S perpendicular to $\overline{b'd'}$. The distance between these supporting lines is clearly w.

It can be shown (Exercise 11.7) that the hexagons $amncpq$ and $a'm'n'c'p'q'$ have the same perimeter. Thus the polygons circumscribed about O and S that result from the polygons of 2^n sides by replacing $\overline{be}, \overline{bf}, \overline{dg}$ and $\overline{dh}$ by $\overline{em}, \overline{mn}, \overline{nf}$, $\overline{gq}, \overline{qp}, \overline{ph}$ (and a similar replacement of $\overline{b'e'}$, etc.) have the same perimeter. Upon carrying out this construction at each vertex, we obtain equiangular polygons of 2^{n+1} sides and conclude that their perimeters are equal.

Thus the given polygons of 2^n sides ($n = 2, 3, 4, \ldots$) circumscribed about O and S have the same perimeters. Therefore, the limits of these perimeter as $n \to \infty$ are equal. Since the perimeter of the circle is πw, the perimeter of S must also be πw. ∎

Although the perimeters of sets of constant width w are all equal, their areas certainly are not. The upper and lower bounds on the area are described in the following theorem, originally formulated by Blaschke.

11.12. Theorem. Among all the sets of constant width w, the circle of diameter w has the greatest area and the Reuleaux triangle of width w has the least.

PROOF. Exercise 11.11. ∎

SETS OF CONSTANT WIDTH

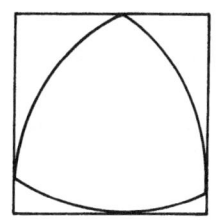

Figure 11.8.

This section closes with two physical applications of sets of constant width.

11.13. Application. If we draw two pairs of parallel supporting lines to a curve of constant width w so that the lines of the one pair are perpendicular to the lines of the other, then we obtain a square of side w. (See Figure 11.8.) Since this can be done with any desired orientation to the square, it follows that a set of constant width can be turned inside a square so that it maintains contact with all four sides. Thus a drill shaped like a Reuleaux triangle can be used to bore a hole with straight sides. The center of the triangle moves in an eccentric path, and the corners of the triangle are the cutting edges.

11.14. Application. In order to project a movie film, the film must make a brief, quick movement (with the shutter closed), followed by a momentary pause (with the shutter open). The gear mechanism that provides this intermittent motion is based on the Reuleaux triangle.

Let P be a plate that is free to move horizontally but not vertically. (See Figure 11.9.) Cut a rectangular hole in the center of P and place a Reuleaux triangle T of width w in it. If the triangle is rotated continuously about one corner x, the plate P will move back and forth intermittently. As T rotates through 120° from position (a) to (b), the plate will move to the right. As T rotates through the next 60° the plate will remain still. During the next 120° turn the plate will move back to the left, again followed by a pause while T finishes the last 60° of its complete revolution. Thus the continual rotation of T is changed into the intermittent linear motion of P.

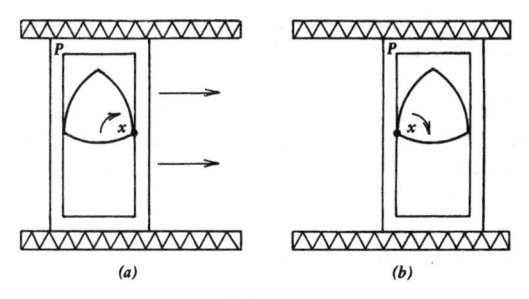

Figure 11.9.

EXERCISES

11.1. Show that the perimeter of a Reuleaux triangle of width w is equal to πw.

11.2. Find the measure of the interior angle at a corner of a Reuleaux triangle.

11.3. Find the area of a Reuleaux triangle of width w.

11.4. Prove Corollary 11.4.

11.5. Find an example of a compact convex subset of E^2 that has more than one incircle.

11.6. Prove that the circumcircle of a compact convex subset of E^2 is unique.

11.7. Let $abcd$ be a rhombus and let $\overline{mn}$ and $\overline{pq}$ be two chords perpendicular to the diagonal $\overline{bd}$ whose distance apart is w. Prove that the perimeter of the hexagon $amncqp$ does not depend on the position of $\overline{mn}$ and $\overline{pq}$.

11.8. Use Exercise 5.8 to prove the following: Let A and B be nonempty compact subsets of E^2 and let ℓ be a line in E^2. Suppose A has width w_1 in the direction of ℓ and B has width w_2 in the direction of ℓ. Then $A + B$ has width $w_1 + w_2$ in the direction of ℓ.

11.9. Use Exercise 11.8 to prove the following: Suppose A and B are sets of constant width w_1 and w_2, respectively. Then $A + B$ has constant width $w_1 + w_2$.

11.10. A nonempty closed convex set S is said to be **strictly convex** if $\overline{xy} \subset \text{bd } S$ implies $x = y$. Prove that every nonempty closed convex set of constant width is strictly convex.

*11.11. Prove Theorem 11.12.

*11.12. Prove the converse to Theorem 11.6: Let S be a nonempty compact convex subset of E^2. Suppose that the diameter of $S \cup \{x\}$ is greater than the diameter of S whenever $x \notin S$. Then S has constant width.

*11.13. Prove that a semicircle of diameter d can be inscribed into any set of constant width d.

*11.14. Let K be the boundary of a compact convex set S in E^2. Suppose that no rectangle has exactly three of its vertices in K. Prove that S is a circle.

*11.15. Let S be an n-dimensional simplex. Denote by R the radius of the smallest sphere containing S and by r the radius of the largest sphere contained in S. That is, R is the circumradius and r is the inradius of S. Prove that $R \geq nr$.

SECTION 12. UNIVERSAL COVERS

Section 11 introduced the concept of the diameter of a set. Closely related to it is the idea of a universal cover.

UNIVERSAL COVERS

12.1. Definition. A compact subset K of E^2 is a **universal cover** in E^2 if any subset S of E^2 having diameter one can be covered by a congruent copy of K.

Since the diameter of a set S is the same as the diameter of cl(conv$S$), it follows that we need only consider compact convex sets S when verifying that K is a universal cover. (See Exercise 12.1.) The problem of finding the smallest universal cover having a given shape has been of interest to mathematicians for many years. The solution to the problem for squares and circles (disks) can be obtained as an application of a Helly-type theorem.

12.2. Theorem. The smallest square that is a universal cover in E^2 has sides of length one.

PROOF. Let S be a subset of E^2 having diameter 1 and let K be a square of side length 1. To prove that K is a universal cover, it suffices by Theorem 6.5 to show that given any three points, say x_1, x_2, x_3, of S, there exists a translate of K that covers $\{x_1, x_2, x_3\}$. Now since the diameter of S is 1, it follows that $\{x_1, x_2, x_3\}$ is contained in a Reuleaux triangle T of width 1. But a Reuleaux triangle of width 1 can be rotated through $360°$ inside a square of side 1. Therefore, no matter what the orientation of T and K, there must be a translate of K that covers T and thus also covers $\{x_1, x_2, x_3\}$. Clearly, no smaller square can be a universal cover since it would not cover a circle of diameter 1. ∎

Actually in Theorem 12.2 we have proved a slightly stronger result than was claimed. We have shown that the square K of side 1 is a *strong* universal cover (i.e., that any set of diameter 1 can be covered by a translate of K and not merely a congruent copy of K).

12.3. Theorem. The smallest circle that is a universal cover in E^2 has radius $1/\sqrt{3}$.

PROOF. The proof that a circle of radius $1/\sqrt{3}$ is a universal cover in E^2 is similar to that of Theorem 12.2 and is left to the reader (Exercise 12.2). No smaller circle can be a universal cover since it would not cover an equilateral triangle of side 1. ∎

The smallest regular hexagonal universal cover is found as an interesting application of the Intermediate Value Theorem from elementary calculus. This in turn leads us to find the smallest triangular universal cover.

12.4. Theorem. The smallest regular hexagon which is a universal cover in E^2 has sides of length $1/\sqrt{3}$.

PROOF. Let S be a set of diameter 1. By Exercise 12.5 we may assume that S is of constant width 1. We begin by circumscribing a rhombus, say *abcd*, about S so that the angle at a is $60°$. (See Figure 12.1.)

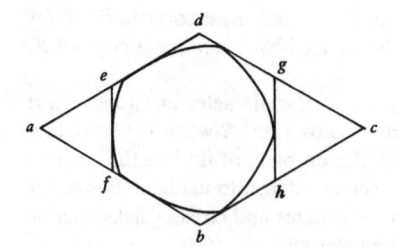

Figure 12.1.

The two support lines perpendicular to the diagonal $\overline{ac}$ of the rhombus cut off two triangles, say *aef* and *cgh*, from *abcd*. If $\overline{ef}$ has the same length as $\overline{gh}$, then the hexagon *efbhgd* is regular and circumscribes S. If not, then we claim that some rotation of the rhombus about S will yield the desired regular hexagon.

If $a'b'c'd'$ is another circumscribed rhombus about S, let α be the counterclockwise angle beteen the diagonal $\overline{ac}$ and the diagonal $\overline{a'c'}$. Let $m(\alpha)$ denote the length of the support line $\overline{e'f'}$ and $n(\alpha)$ the length of the support line $\overline{g'h'}$. Then $m(\alpha)$ and $n(\alpha)$ are continuous functions of α.

If $\overline{ef}$ and $\overline{gh}$ are not the same length, then $m(0) \neq n(0)$, and we may assume without loss of generality that $m(0) - n(0) > 0$. But the rhombus corresponding to $\alpha = \pi$ will yield the same hexagon as *efbhgd*, except that the sides $\overline{ef}$ and $\overline{gh}$ will be interchanged. Thus $m(\pi) - n(\pi) < 0$. Since the difference $m(\alpha) - n(\alpha)$ is a continuous function of α, it follows from the Intermediate Value Theorem that there exists an α between 0 and π such that $m(\alpha) - n(\alpha) = 0$. For this α, the corresponding circumscribing hexagon will be regular.

Since S has diameter 1, its maximum width is 1 (Theorem 11.3), and so the regular hexagon circumscribed about S will have sides of length at most $1/\sqrt{3}$. No smaller regular hexagon can be a universal cover since it would not cover a circle of diameter 1. ∎

12.5. Theorem. *The smallest equilateral triangle that is a universal cover in* E^2 *has sides of length* $\sqrt{3}$.

PROOF. An equilateral triangle with sides of length $\sqrt{3}$ is a universal cover in E^2 since it can cover a regular hexagon of side length $1/\sqrt{3}$, and such a hexagon is itself a universal cover by Theorem 12.4. That no smaller such triangle is a universal cover again follows from the circle of diameter 1. ∎

Having found several "smallest" universal covers in E^2, we are led to pose the following question: For each fixed integer k, what is the smallest regular k-gon that is a universal cover in E^2? We have shown that for $k = 3$, 4, and 6 the answer is the k-gon circumscribed about a circle of diameter 1. The answer for k other than these is not known at this time (1982). We might be tempted to conjecture that any regular k-gon circumscribed about a circle of diameter 1

EXERCISES

would be a universal cover in E^2. This, however, is not true for any $k > 6$. (See Exercise 12.3.) Indeed, as k increases, the circumscribed k-gons approach the circle, and the smallest circular universal cover was found to have diameter $2/\sqrt{3}$.

The most famous unsolved problem dealing with universal covers is the following question posed by Lebesgue:

What is the minimum area A that a universal cover in E^2 can have?

Since any universal cover must be as large as the circle of diameter 1, A must be greater than or equal to $\pi/4$. [In fact it can be shown (Exercise 12.5) that every set of diameter 1 is contained in some set of constant width 1. It follows then from Theorem 11.12 that among all sets of diameter 1, the circle has the largest area.] Since the regular hexagon of side length $1/\sqrt{3}$ is a universal cover, A must be less than or equal to $\sqrt{3}/2$. Hence we know that A satisfies the inequality

$$0.785 \approx \frac{\pi}{4} \leq A \leq \frac{\sqrt{3}}{2} \approx 0.866,$$

but the exact value of A is not known. (See Exercise 12.10 for even sharper bounds.)

EXERCISES

12.1. (a) Let S be a compact subset of E^2. Prove that the diameter of S is equal to the diameter of conv S.
  (b) Let S be a bounded subset of E^2. Prove that the diameter of S is equal to the diameter of cl S.

12.2. Prove Theorem 12.3.

12.3. Let K be a regular k-gon ($k > 6$) circumscribed about a circle of diameter 1. Find a set S of diameter 1 which cannot be covered by a congruent copy of K.

12.4. Find what is wrong with the following "proof": Let S have diameter 1. Then every three points of S can be covered by a Reuleaux triangle of width 1. By Theorem 6.5, there exists a Reuleaux triangle of width 1 that covers S. Thus a Reuleaux triangle of width 1 is a universal cover in E^2.

*****12.5.** Prove that every subset of E^2 that has diameter w is contained in a set of constant width w.

12.6. Find what is wrong with the following "proof": Let S have diameter 1. By Exercise 12.5 there exists a set S^* containing S that has constant width 1. Let T and T' be two squares of side length 1 circumscribed about S^* such that the orientation of T' is a 45° rotation of T. Then $T \cap T'$ is a regular octagon of side $\sqrt{2} - 1$

circumscribing S^*. Thus the regular octagon of side length $\sqrt{2} - 1$, that is, the regular octagon circumscribed about a circle of diameter 1, is a universal cover in E^2.

12.7. Let S be a compact subset of E^2 having diameter 1. Prove that S is a union of three sets each of diameter less than 1.

12.8. Prove the following theorem of D. Gale: Let S be a compact subset of E^2 having diameter d. Then S is a union of three sets each of diameter less than or equal to $\sqrt{3}\,d/2$.

†12.9. Prove or disprove the following conjecture of K. Borsuk: Every subset of E^n whose diameter is d can be expressed as the union of $n + 1$ sets each of diameter less than d.

*12.10. (a) Prove that the minimal area A of a universal cover in E^2 must satisfy

$$0.8257 \approx \frac{\pi}{8} + \frac{\sqrt{3}}{4} \leq A \leq \frac{2(3 - \sqrt{3})}{3} \approx 0.8454.$$

(b) Prove that the minimal perimeter P of a universal cover in E^2 must satisfy

$$3.302 \approx \sqrt{3} + \frac{\pi}{2} \leq P \leq 8 - \frac{8}{\sqrt{3}} \approx 3.382.$$

SECTION 13. THE ISOPERIMETRIC PROBLEM

Section 12 introduced Lebesgue's problem of finding a universal cover in E^2 of minimal area. This is related to a whole class of "extremum" problems wherein one quantity is to be minimized (or maximized) subject to certain restraining conditions. The best known of all the extremum problems is the classical isoperimetric problem:

Among all plane figures of perimeter 1, find the figure with the greatest area.

Using the techniques of calculus it is easy to show, for example, that among all the rectangles of perimeter 1, the square has the greatest area. Likewise, among all triangles of perimeter 1, the equilateral triangle has the greatest area. The isoperimetric problem cannot, however, be solved by the techniques of beginning calculus since no restrictions are placed on the shape of the figures involved (except that they have a perimeter of length 1).

The isoperimetric problem as just stated says nothing about the sets being convex. It is difficult (indeed, impossible) to define the area and perimeter of an arbitrary subset of the plane. Section 11 gave a precise definition of the area and perimeter of any planar compact convex set having an interior, but to

THE ISOPERIMETRIC PROBLEM

extend the definitions to more general sets would be considerably beyond the scope of this book.

Fortunately, it is not at all unreasonable to expect that a solution to the isoperimetric problem would be a compact convex set having an interior. If we assume, for the moment, an intuitive notion of the area and perimeter of a nonconvex set S, then the convex hull of S will have a shorter perimeter and a larger area. (See Figure 13.1.) It is also clear that a figure of fixed perimeter and maximum area would need to be compact and have an interior. Thus in our present discussion we shall assume that any solution to the isoperimetric problem will have these properties.

Since for two similar figures in the plane the ratio of the area to the square of the perimeter is the same for both, the isoperimetric problem can be rephrased as a minimizing problem:

Among all plane figures of area 1, find the figure with the minimum perimeter.

Clearly a solution to one problem entails a solution to the other. Specifically, if S is a solution to the original maximizing problem, then a set similar to S will be a solution to the minimizing problem, and vice versa.

Unlike Lebesgue's problem on universal covers, the isoperimetric problem has been solved. The first purported solution was formulated in 1838 by the Swiss geometer Jakob Steiner. He proved that given any noncircular convex figure B there can be found a figure B' with the same perimeter but greater area. At first glance this would seem to prove that the circle of perimeter 1 has the largest area and is, therefore, the solution to the isoperimetric problem. That this is not the case is best illustrated by an example:

Given any positive integer k other than 1, we can find another positive integer $k' = k^2$ that is greater than k.

It is, of course, incorrect to conclude from this that 1 is the largest positive integer. Proving that no integer different from 1 is the largest does not prove that 1 is the largest since, in fact, a largest integer does not exist.

The necessity of proving the existence of a set with perimeter 1 and maximum area is precisely the weakness in Steiner's solution to the isoperimetric problem. Having said this, we now proceed to give one of Steiner's proofs that a noncircular set cannot be maximal. (He actually proved this in five different ways.) We shall return to the question of existence in Section 16 and

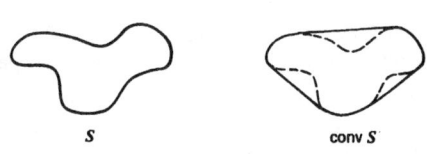

Figure 13.1.

SPECIAL TOPICS IN E²

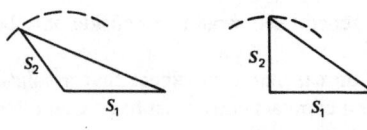

Figure 13.2.

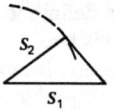

ultimately complete our argument by using approximating polytopes in Section 22. Assuming for the moment these later results, we can conclude that the circle of perimeter 1 is indeed the unique solution to the isoperimetric problem. We begin with two lemmas and a definition.

13.1. Lemma. Among all the triangles with two given sides, the triangle in which the given sides are mutually perpendicular has the greatest area.

PROOF. Let S_1 be one of the given sides and take this side as the base of the triangle. The area of the triangle will be maximal when its height is maximal, and this will occur when the second given side S_2 is perpendicular to the first. (See Figure 13.2.) ∎

13.2. Definition. Let S be a compact convex subset of E^2. A **chord** of S is a line segment between two boundary points of S.

13.3. Lemma. If a chord of a convex set B bisects the perimeter but divides the area into two unequal parts, then there exists a set B' with the same perimeter as B but having a larger area.

PROOF. Let $\overline{xy}$ be a chord of the convex set B that bisects the perimeter but that divides B into two parts B_1 and B_2 with the area of B_1 greater than the area of B_2. (See Figure 13.3.) Let B_1^* be the reflection of B_1 in the line $\overline{xy}$. Then the set $B' = B_1 \cup B_1^*$ has the same perimeter as B but a larger area. ∎

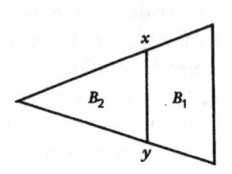

Figure 13.3.

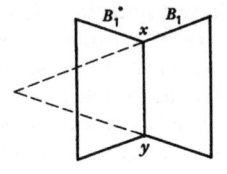

13.4. Theorem. Let S be a compact convex subset of E^2 with int $S \neq \emptyset$. If S is not a circle, then there exists a set S' that has the same perimeter as S but greater area.

PROOF. Let x and y be points on the boundary of S that bisect the perimeter. If the chord $\overline{xy}$ divides the area of S into two unequal parts, then

THE ISOPERIMETRIC PROBLEM

Lemma 13.3 implies the existence of the desired set S'. Thus we may assume that $\overline{xy}$ divides S into two parts S_1 and S_2 that have equal area.

If S is not a circle, there must be at least one boundary point p of S such that $\angle xpy$ is not a right angle, and we may assume that $p \in S_1$. The segments $\overline{xp}$ and $\overline{py}$ divide S_1 into three parts: the triangle $\triangle xpy$, the sector A_1 cut off by the chord $\overline{xp}$, and the sector A_2 cut off by the chord $\overline{py}$. (See Figure 13.4.) We now replace S_1 with a new figure S_1^* by leaving A_1 and A_2 unchanged and replacing $\angle xpy$ by a right angle $\angle x'p'y'$. Finally let S' be the union of S_1^* and the reflection of S_1^* in the chord $\overline{x'y'}$. Then S' has the same perimeter as S and it follows from Lemma 13.1 that the area of S' is greater than the area of S. ∎

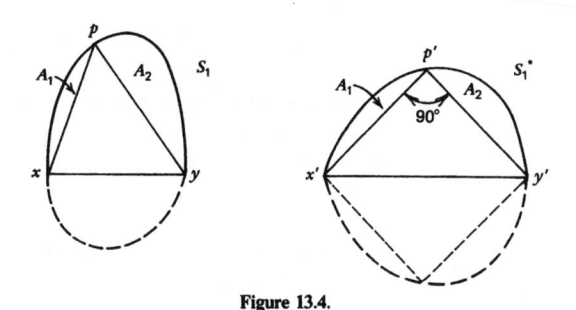

Figure 13.4.

The isoperimetric property of the circle also can be stated algebraically. For a circle, the square of the perimeter is equal to 4π times its area. Thus we have the following so called isoperimetric inequality for any convex set S:

$$p^2 \geq 4\pi A,$$

where p is the perimeter and A is the area of S. Furthermore, equality holds in this expression iff S is a circle.

We conclude this section with a simple corollary of the isoperimetric problem which is referred to as "the problem of Dido." According to legend, Dido, the mythical queen of Phoenicia, made an agreement with a North African tribe that a piece of land bounded by an ox hide would be given to her. But instead of covering a small plot of ground as the tribe expected, she cut the hide into narrow strips and tied these together into a long rope. To further her advantage she used the (straight) seacoast as one of the boundaries. Thus instead of the isoperimetric problem, Dido was faced with the following:

> Among all the plane figures bounded by a segment of arbitrary length and by a curve of length ℓ, find the figure with the greatest area.

The answer to Dido's problem is the following theorem.

13.5. Theorem. Of all the plane figures that are bounded by a segment of arbitrary length and by a curve of length ℓ, the semicircle of radius ℓ/π has the greatest area.

PROOF. Let S be a semicircle of radius ℓ/π and let B be any other figure bounded by a line segment and a curve of length ℓ. We reflect the two sets in the lines on which the line segments in their boundaries lie. Thus we obtain a circle S' and another set B' having the same perimeter as S'. By the isoperimetric problem, S' has greater area than B', and so S (which is half of S') has greater area than B (which is half of B'). ∎

EXERCISES

Note: You may assume the existence of any maximal or minimal figures required in the following exercises.

- **13.1.** Prove that among all triangles with a fixed base and a fixed perimeter, the isosceles triangle has maximum area.
- **13.2.** Prove that among all triangles with equal perimeter, the equilateral triangle has maximum area.
- **13.3.** Prove that among all polygons whose corresponding sides have equal lengths, the polygon inscribed in a circle has maximum area.
- **13.4.** Prove that among all k-gons inscribed in a given circle, the regular k-gon has maximum area.
- ***13.5.** Prove that among all k-gons with equal perimeter, the regular k-gon has maximum area.
- **†13.6.** What is the largest area that a k-gon of unit diameter can have?
- **†13.7.** Any k points in a unit square determine a largest convex area within the square containing none of the k points. Determine $f(k)$, the minimum (of the largest areas) taken over all sets of k points.
- **†13.8.** The isoperimetric problem can be extended to include figures lying on two-dimensional surfaces in E^3. Classify the surfaces in E^3 for which the isoperimetric inequality $p^2 \geq 4\pi A$ remains valid.
- **†13.9.** Let S be a compact convex subset of E^2. A point p in S is called an **equichordal point** of S if each chord of S through p has the same length. Can S have more than one equichordal point?
- ***13.10.** Let S be a compact convex subset of E^2 having constant width. If S has an equichordal point, prove that S is a circle.
- ***13.11.** Let S be a compact convex subset of E^2. A point p in S is called a **center** of S if each chord of S through p is bisected by p. Prove that the only set of constant width that has a center is a circle.
- **13.12.** *(a) Let K be a compact convex subset of E^2 with int $K \neq \emptyset$. Let P be a convex m-gon ($m \geq 3$) of minimum area containing K. Prove that the midpoints of the sides of P lie on the boundary of K.

EXERCISES

*(b) Let K be a parallelogram in E^2 and let T be any triangle containing K. Prove that area $(T) \geq 2$ area (K).

*(c) Let K be a compact convex subset of E^2 with int $K \neq \varnothing$, and let T be a triangle of minimum area that contains K. Prove that area $(T) \leq 2$ area (K). Furthermore, if area $(T) = 2$ area (K), then K must be a parallelogram.

*(d) Let K be a compact convex subset of E^2 with int $K \neq \varnothing$. Prove that K is contained in a quadrilateral Q such that area $(Q) \leq \sqrt{2}$ area (K).

†(e) Is part (d) the best possible result? That is, does there exist a compact convex subset K of E^2 with int $K \neq \varnothing$ such that all of its circumscribing quadrilaterals have area at least $\sqrt{2}$ area (K)?

6
FAMILIES OF CONVEX SETS

In many extremal geometrical problems, being able to prove the existence of a solution is essential. Once a solution is known to exist, the problem of finding it is often simplified. In this connection, it is useful to consider certain families of convex sets and to be able to assert that within the family it is possible to select a particular set having the desired properties.

The main objective of this chapter is to prove the Blaschke selection theorem. This theorem enables us to conclude that the family of closed convex subsets of a compact convex set in E^n can be made into a compact metric space. Then, since any real-valued continuous function attains its maximum and minimum on a compact set, we can establish the existence of sets having certain extremal properties.

SECTION 14. PARALLEL BODIES

In order to make a family of convex sets into a metric space, we need to have a measure of the "distance" between two subsets A and B of E^n. We have already introduced (Exercises 1.17 and 1.18) one such notion of distance:

$$d(A, B) = \inf\{d(x, y): x \in A \text{ and } y \in B\}.$$

This earlier definition proves to be inadequate for our present purposes since it fails to discriminate sufficiently between sets. We would like the distance between two sets to be small only if the two sets are nearly the same, both in shape and position. Our earlier definition only measures their closeness in position. For a more restrictive definition of distance, the following concept is useful.

14.1. Definition. Let A be a nonempty subset of E^n. The **parallel body** A_δ is defined to be

$$A_\delta \equiv \bigcup_{a \in A} K(a, \delta),$$

where $K(a, \delta) \equiv \{x: d(x, a) \leq \delta\}$ and $\delta > 0$.

PARALLEL BODIES

It follows readily from the definition that A_δ is equal to $A + K(\theta,\delta)$. Thus when A is convex, so is A_δ, since it is the sum of two convex sets. We can also describe A_δ by $A_\delta = \{x: d(x, A) \leq \delta\}$. This latter characterization is often the easiest to use in applications.

14.2. Theorem. If A is a nonempty subset of E^n, then

$$(A_\delta)_\varepsilon = A_{\delta + \varepsilon}.$$

PROOF. Exercise 14.1. ∎

14.3. Definition. Let A and B be nonempty compact subsets of E^n. Then the **distance D between A and B** is defined as

$$D(A, B) \equiv \inf\{\delta: A \subset B_\delta \text{ and } B \subset A_\delta\}.$$

14.4. Example. let $A = \{(x, y): 1 \leq x \leq 3 \text{ and } 1 \leq y \leq 3\}$ and let $B = \{(x, y): (x - 4)^2 + (y - 1)^2 \leq 1\}$. Then $d(A, B) = 0$ and $D(A, B) = \sqrt{13} - 1$. (See Figure 14.1.)

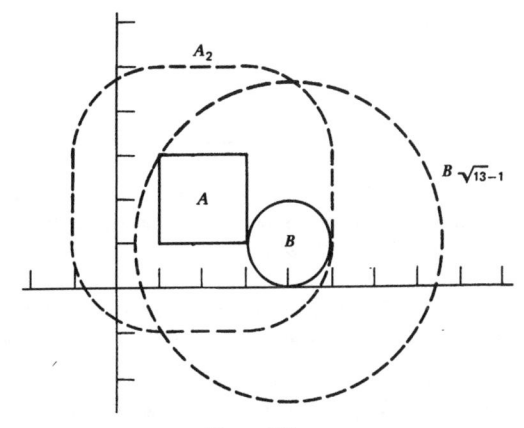

Figure 14.1.

14.5. Theorem. The distance function D has the following four properties for any nonempty compact subsets A, B, C of E^n:
  (a) $D(A, B) > 0$ if $A \neq B$.
  (b) $D(A, B) = 0$ if $A = B$.
  (c) $D(A, B) = D(B, A)$.
  (d) $D(A, C) \leq D(A, B) + D(B, C)$.

PROOF. Properties (a), (b), and (c) follow directly from the definition. To prove (d), let

$$\alpha = D(A, B) \quad \text{and} \quad \beta = D(B, C),$$

and let $\gamma = \alpha + \beta$. Since $B \subset A_\alpha$, Theorem 14.2 implies

$$B_\beta \subset (A_\alpha)_\beta = A_{\alpha+\beta} = A_\gamma.$$

Thus $C \subset B_\beta \subset A_\gamma$. Similarly, $B \subset C_\beta$, so that $A \subset B_\alpha \subset C_{\alpha+\beta} = C_\gamma$. Therefore, we have $C \subset A_\gamma$ and $A \subset C_\gamma$ so that $D(A, C) \leq \gamma = \alpha + \beta = D(A, B) + D(B, C)$, as was to be shown. ∎

A function such as D that satisfies the four properties given in Theorem 14.5 is called a **metric**. Thus D is a metric on the collection $\mathcal{C}$ of all nonempty compact convex subsets of E^n. (Actually, D is a special case of the so-called Hausdorff metric for compact sets.) Knowing that D makes $\mathcal{C}$ into a metric space enables us to fully exploit the distance properties of D. Specifically, it enables us to consider the convergence of sequences of convex sets.

14.6. Definition. A sequence $\{A_i\}$ of compact convex subsets of E^n is said to **converge** to a set A if

$$\lim_{i \to \infty} D(A_i, A) = 0.$$

We also say that A is the **limit** of the sequence $\{A_i\}$.

The usefulness of the metric D in working with convex sets is enhanced greatly by the following theorem, which asserts that the limit of a sequence of convex sets must also be a convex set.

14.7. Theorem. *Let $\{A_i\}$ be a sequence of compact convex subsets of E^n and suppose that $\{A_i\}$ converges to a compact set A. Then A is also convex.*

PROOF. If A is not convex, then there exist points x and y in A and a point z not in A such that $z \in \overline{xy}$. (See Figure 14.2.) Since A is closed, there exists a positive number δ such that

$$B(z, \delta) \cap A = \varnothing.$$

Choose i sufficiently large so that $D(A_i, A) < \delta/2$ and select two points x' and y' in A_i such that

$$d(x, x') < \frac{\delta}{2} \quad \text{and} \quad d(y, y') < \frac{\delta}{2}.$$

THE BLASCHKE SELECTION THEOREM

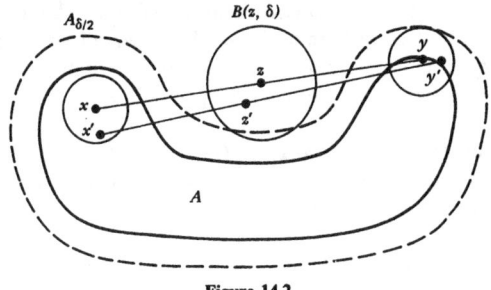

Figure 14.2.

It follows that there exists a point z' in $\overline{x'y'}$ such that

$$d(z, z') < \frac{\delta}{2}.$$

But then $d(z', A) > \delta/2$ so that $z' \notin A_{\delta/2} \supset A_i$. This contradicts the convexity of A_i and establishes that A must be convex. ∎

EXERCISES

14.1. Prove Theorem 14.2.

14.2. Consider the following three subsets of $\mathbf{E}^2$:
$A = \{(x, y): -1 \leq x \leq 2 \text{ and } 2 \leq y \leq 3\}$,
$B = \{(x, y): 1 \leq x \leq 2 \text{ and } -1 \leq y \leq 1\}$, and
$C = \{(x, y): (x + 1)^2 + y^2 \leq 1\}$.
Find $D(A, B)$, $D(A, C)$, and $D(B, C)$.

14.3. Find an example of three nonempty compact convex subsets A, B, and C of $\mathbf{E}^2$ such that $A \subset B \subset C$ with $A \neq B \neq C$, but $D(A, B) = D(B, C) = D(A, C)$.

14.4. Let $S = \text{cl } B(x, \delta)$ and let p be any point in $\mathbf{E}^n$. Find $D(\{p\}, S)$.

14.5. Let $S_1 = \text{cl } B(x, \alpha)$ and $S_2 = \text{cl } B(y, \beta)$. Find $D(S_1, S_2)$.

SECTION 15. THE BLASCHKE SELECTION THEOREM

Having made the collection $\mathcal{C}$ of all nonempty compact convex subsets of $\mathbf{E}^n$ into a metric space using the distance function D, we now need a criterion for determining when a subcollection of $\mathcal{C}$ is compact. In this regard, the following definition is very useful.

15.1. Definition. A subcollection $\mathfrak{M}$ of $\mathcal{C}$ is **uniformly bounded** if there exists a solid sphere in $\mathbf{E}^n$ that contains all the members of $\mathfrak{M}$.

The Blaschke selection theorem (also called the Blaschke convergence theorem) asserts that an infinite uniformly bounded collection in $\mathcal{C}$ contains a convergent sequence. It performs a role for families of sets in $\mathcal{C}$ similar to what the Bolzano-Weierstrass theorem does for points in $\mathbf{E}^n$.

15.2. Theorem (Blaschke Selection Theorem). Let $\mathcal{M}$ be a uniformly bounded infinite subcollection of $\mathcal{C}$. Then $\mathcal{M}$ contains a sequence of distinct terms that converges to a member of $\mathcal{C}$.

The proof of Theorem 15.2 will be divided into two main parts, each part being called a lemma.

15.3. Lemma. Let $\mathcal{M}$ be a uniformly bounded infinite subcollection of $\mathcal{C}$. Then $\mathcal{M}$ contains a **Cauchy sequence** $\{C_k\}$ of distinct terms. That is, the sequence $\{C_k\}$ has the following property: given any $\varepsilon > 0$, there exists an integer N such that

$$0 < D(C_i, C_j) < \varepsilon \quad \text{whenever } i, j > N \text{ and } i \neq j.$$

PROOF. Using the standard basis for $\mathbf{E}^n$, let A be a box (cf. Definition 10.1) which contains each of the members of $\mathcal{M}$ and such that the distance between any two parallel sides is τ. By subdividing each edge of A into 2^i equal parts we can partition A into congruent closed boxes, each of which has edges of length $2^{-i}\tau$. The set of congruent boxes thus obtained will be denoted by K_i ($i = 1, 2, \ldots$) and the union T of a subset of K_i will be called a minimal covering for a set M in $\mathcal{M}$ if $M \subset T$ and if M intersects each box belonging to T. (See Figure 15.1.)

Since K_1 consists of a *finite* number of boxes (2^n to be precise), there are only finitely many possible minimal coverings from K_1. Since $\mathcal{M}$ has an infinite number of members, there must exist a countably infinite sequence of members of $\mathcal{M}$, denoted by $\{C_{1\alpha}: \alpha = 1, 2, \ldots\}$, which have the same minimal covering from K_1. Similarly, the sequence $\{C_{1\alpha}\}$ contains an infinite subsequence $\{C_{2\alpha}: \alpha = 1, 2, \ldots\}$ whose members all have the same minimal covering from K_2. Proceeding in this way, we obtain a double sequence $\{C_{i\alpha}: i, \alpha = 1, 2, \ldots\}$ such that for each fixed i, the sets $C_{i\alpha}$ ($\alpha = 1, 2, \ldots$) have the same minimal covering.

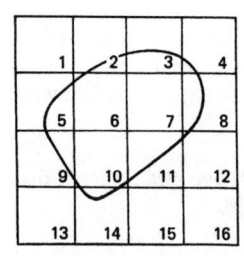

Figure 15.1. The minimal covering for M is the union of the boxes numbered 2, 3, 4, 5, 6, 7, 8, 9, 10, 11, 12, 14.

THE BLASCHKE SELECTION THEOREM

Now if two sets have the same minimal covering, then the distance between them cannot be greater than the diagonal length of each box in the covering. Thus for each fixed i, we have

$$D(C_{i\alpha}, C_{i\beta}) \leq \frac{\tau\sqrt{n}}{2^i}.$$

Furthermore, since $\{C_{j\beta}: \beta = 1, 2, \dots\}$ is a subsequence of $\{C_{i\alpha}: \alpha = 1, 2, \dots\}$ when $j > i$, we have

$$D(C_{i\alpha}, C_{j\beta}) \leq \frac{\tau\sqrt{n}}{2^i}$$

for all α and β when $j \geq i$.

Finally, let $C_k \equiv C_{kk}$ ($k = 1, 2, \dots$) be the diagonal sequence. Then for all $j \geq i$ we have $D(C_i, C_j) \leq \tau\sqrt{n}/2^i$. Thus, given any positive ε, there exists an integer N such that $D(C_i, C_j) < \varepsilon$ whenever $i, j > N$. ∎

15.4. Lemma. Let $\mathfrak{M}$ be a uniformly bounded subcollection of $\mathcal{C}$. If $\{C_k\}$ is a Cauchy sequence in $\mathfrak{M}$, then there exists a nonempty compact convex set B such that $\{C_k\}$ converges to B.

PROOF. Let $B_k \equiv \text{cl}(C_k \cup C_{k+1} \cup C_{k+2} \cup \cdots)$. Thus for each k, $B_{k+1} \subset B_k$. Then let

$$B \equiv B_1 \cap B_2 \cap B_3 \cap \cdots = \bigcap_{k=1}^{\infty} B_k.$$

Since $\{B_k\}$ is a decreasing nested sequence of nonempty compact sets, B must be a nonempty compact set (Exercise 15.1). Let $(B)_\varepsilon$ and $(C_k)_\varepsilon$ denote the parallel bodies to B and C_k.

We claim that given any positive ε, there exists an integer N_1 such that $B_k \subset \text{int}(B)_\varepsilon$ for all $k > N_1$. Indeed, if this were not the case, then

$$A_k \equiv B_k \cap \text{bd}(B)_\varepsilon \qquad k = 1, 2, \dots$$

would form a decreasing nested sequence of nonempty compact sets, and there would exist a point x in $\bigcap_{k=1}^{\infty} A_k$. But then x would be in $\bigcap_{k=1}^{\infty} B_k$ and not in B, a contradiction. Therefore, there must exist an integer N_1 such that $B_k \subset \text{int}(B)_\varepsilon$ for all $k > N_1$. Since $C_i \subset B_k$ for all $i \geq k$, this implies that

$$C_k \subset \text{int}(B)_\varepsilon \subset (B)_\varepsilon \qquad \text{for } k > N_1.$$

On the other hand, given any positive ε there exists an integer N_2 such that

$$D(C_i, C_j) < \frac{\varepsilon}{2} \qquad \text{whenever } i, j > N_2,$$

since $\{C_k\}$ is Cauchy. Thus for each $k > N_2$, we have

$$\bigcup_{j=k}^{\infty} C_j \subset (C_k)_{\varepsilon/2},$$

and so

$$B \subset B_k = \text{cl}\left(\bigcup_{j=k}^{\infty} C_j\right) \subset (C_k)_{\varepsilon}.$$

Finally, combining the preceding results, we have

$$D(C_k, B) < \varepsilon \quad \text{when } k > \max\{N_1, N_2\}.$$

Thus $\{C_k\}$ converges to the nonempty compact set B. By Theorem 14.7 we know that B must also be convex. ∎

We are now in a position to combine Lemmas 15.3 and 15.4 and finish the proof of Blaschke's selection theorem.

PROOF OF THEOREM 15.2. If $\mathfrak{M}$ is a uniformly bounded infinite subcollection of $\mathcal{C}$, then by Lemma 15.3, $\mathfrak{M}$ contains a Cauchy sequence $\{C_k\}$, and by Lemma 15.4 this Cauchy sequence must converge to a member of $\mathcal{C}$. ∎

In view of the Blaschke selection theorem we make the following definition of compactness in $\mathcal{C}$. It can be shown (Exercise 15.2) that this definition (also called sequential compactness) is equivalent to the "closed and bounded" definition given earlier when $\mathcal{C}$ is regarded as a metric space with metric D.

15.5. Definition. A family $\mathfrak{F}$ of convex sets in $\mathcal{C}$ is a **compact family** if every sequence from $\mathfrak{F}$ has a convergent subsequence whose limit belongs to $\mathfrak{F}$.

The usefulness of this formulation of compactness is made evident in the following theorem. (See Definition 16.3.)

15.6. Theorem. Let $\mathfrak{F}$ be a nonempty compact family in $\mathcal{C}$ and suppose g is a continuous real-valued function defined on the members of $\mathfrak{F}$. Then g attains its maximum and minimum values on $\mathfrak{F}$.

PROOF. We will only prove that g attains its maximum on $\mathfrak{F}$, since the proof for the minimum is similar. To this end, we observe that the set $\{g(F): F \in \mathfrak{F}\}$ is a bounded subset of E^1. Indeed, if it were not, then we could find a sequence $\{F_i\}$ in $\mathfrak{F}$ such that $g(F_i) \to \infty$ as $i \to \infty$. The compactness of $\mathfrak{F}$ would then imply the existence of a subsequence of $\{F_i\}$ that converges to a set F_0 in $\mathfrak{F}$. The continuity of g implies that $g(F_0) < \infty$ and so the limit of $g(F_i)$ cannot be infinite.

THE EXISTENCE OF EXTREMAL SETS

Having established that $\{g(F): F \in \mathcal{F}\}$ is bounded we may let

$$M = \sup_{F \in \mathcal{F}} g(F).$$

Now let $\{F_i\}$ be a sequence in $\mathcal{F}$ such that $g(F_i)$ converges to M. Since $\mathcal{F}$ is compact, there exists a subsequence of $\{F_i\}$ that converges to a set F' in $\mathcal{F}$. Since g is continuous, we must have $g(F') = M$. ∎

EXERCISES

15.1. Let $\{B_k\}$ be a decreasing nested sequence of nonempty compact subsets of E^n. That is, for each k, $B_{k+1} \subset B_k$. Prove that $B \equiv \bigcap_{k=1}^{\infty} B_k$ is a nonempty compact set.

15.2. Let $\mathcal{F}$ be a family of sets in $\mathcal{C}$. Prove that $\mathcal{F}$ is a compact family if and only if $\mathcal{F}$ is closed and bounded with respect to the metric D.

*15.3. Prove a convergence theorem for a uniformly bounded family of star-shaped sets analogous to the Blaschke Selection Theorem 15.2 for convex sets.

SECTION 16. THE EXISTENCE OF EXTREMAL SETS

On the basis of the theory developed in Section 15, we can prove the existence of convex sets with certain extremal properties. We begin by establishing that every compact convex subset of E^2 has an incircle (see Definition 11.7).

16.1. Theorem. Let $\mathcal{F}$ be a nonempty compact family of solid spheres in E^n. Then $\mathcal{F}$ contains at least one member with a maximum radius.

PROOF. We claim that the radius $r(F)$ of $F \in \mathcal{F}$ is a continuous function defined on $\mathcal{F}$. Indeed, if F_1 and F_2 are concentric spheres, then the difference in their radii (in absolute value) is equal to the distance between them measured by the metric D. If we move F_2 so that it is not concentric with F_1, then the distance between them will increase. Thus for any spheres F_1 and F_2 in $\mathcal{F}$, we have

$$|r(F_1) - r(F_2)| \leq D(F_1, F_2).$$

It follows that r is a continuous function on $\mathcal{F}$.

Therefore, by Theorem 15.6, there exists a sphere F in $\mathcal{F}$ such that $r(F)$ is a maximum. ∎

16.2. Corollary. Let S be a nonempty compact convex subset of E^2. Then there exists a circle of maximum diameter contained in S. That is, S has an incircle.

PROOF. Let $\mathcal{F}$ be the family of all circles contained in S. Then $\mathcal{F}$ is uniformly bounded. If $\{F_i\}$ is a sequence from $\mathcal{F}$, then Blaschke's Theorem 15.2 implies the existence of a subsequence that converges to a compact convex set G. By Exercise 16.1, G is a circle contained in S. Thus $\mathcal{F}$ is a compact family, and by Theorem 16.1 there exists a member of $\mathcal{F}$ whose diameter is a maximum. ∎

We now turn our attention to proving the existence of a solution to the isoperimetric problem (Section 13). First we must define the concept of continuity within the context of a family of compact convex subsets of E^n. It is not difficult to show (Exercise 16.3) that this definition is consistent with our earlier Definition 1.14 when the family is considered as a metric space with the distance given by Definition 14.2.

16.3. Definition. Let $\mathcal{F}$ be a family of nonempty compact convex subsets of E^n. A real-valued function f defined on $\mathcal{F}$ is **continuous** on $\mathcal{F}$ if, given any $S \in \mathcal{F}$ and any sequence $\{S_i\}$ in $\mathcal{F}$ that converges to S, we have $\lim_{i \to \infty} f(S_i) = f(S)$.

The important application of continuity for us at this point is that the area and perimeter functions are both continuous on $\mathcal{F}$. This will be proved in Section 22 (Theorem 22.6), where we use polytopes to approximate the convex sets. Using this result, we obtain the following existence theorem.

16.4. Theorem. Let $\mathcal{F}$ be the family of all plane figures of perimeter 1. Then $\mathcal{F}$ contains a member whose area is a maximum.

PROOF. Let S be circle of radius 1 centered at the origin. Then any member of $\mathcal{F}$ can be translated to be in S. Furthermore, if the interior of a set is empty, then its area is zero, which is certainly not maximal. Also, a nonconvex set cannot have maximal area. Thus we may restrict our attention to the family $\mathcal{F}'$ of all compact convex sets having a nonempty interior that are contained in S and that have perimeter 1.

Since all the members of $\mathcal{F}'$ are contained in S, their areas are bounded above by the area of S. Thus, letting $A(F)$ denote the area of F, we conclude that

$$m \equiv \sup_{F \in \mathcal{F}'} A(F)$$

is a finite number. Let $\{F_i\}$ be a sequence in $\mathcal{F}'$ such that $A(F_i)$ converges to m. Since the F_i are uniformly bounded, by Blaschke's Theorem 15.2 we may select a subsequence $\{F_j\}$ that converges to a compact convex set F. Since the perimeter function $P(F)$ is continuous on $\mathcal{F}'$ (Theorem 22.6),

$$P(F) = P\left(\lim_{j \to \infty} F_j\right) = \lim_{j \to \infty} P(F_j) = 1,$$

EXERCISES

and so $F \in \mathcal{F}'$. Since the area function is also continuous on $\mathcal{F}'$ (Theorem 22.6), we have

$$A(F) = A\left(\lim_{j\to\infty} F_j\right) = \lim_{j\to\infty} A(F_j)$$
$$= \lim_{i\to\infty} A(F_i) = m.$$

Thus the compact convex set F has maximal area. ∎

EXERCISES

16.1. Let $\{F_i\}$ be a sequence of spheres all of which are contained in a compact convex subset S of E^n. Suppose $\{F_i\}$ converges to a compact set G. Prove that G is a sphere contained in S.

16.2. Let S be a nonempty compact convex subset of E^n. Prove that there exists a sphere of minimum diameter containing S. That is, show that S has a circumsphere.

16.3. Show that Definition 16.3 is consistent with Definition 1.14.

*16.4. Let K be a compact convex subset of E^n with int $K \neq \emptyset$. Prove that $R \leq D[n/(2n+2)]^{1/2}$, where R is the circumradius of K and D is its diameter.

†16.5. Let S be a sphere in E^3 and let $\mathcal{F}$ be the family of all compact convex sets contained in S that have a given positive volume. Determine which member of $\mathcal{F}$ has maximum surface area.

7

CHARACTERIZATIONS OF CONVEX SETS

In Theorem 5.4 we obtained an important characterization of closed convex sets having an interior in terms of supporting hyperplanes. We thus were able to equate the global pairwise definition of convexity (the line segment joining *any two* points of S must lie in S) with a property of certain individual points (*each boundary* point must lie on a supporting hyperplane). In this chapter we look at several other ways of characterizing convex sets. In Section 17 we introduce "local convexity," in Section 18 we examine local support properties, and in Section 19 we focus our attention on nearest-point properties.

SECTION 17. LOCAL CONVEXITY

One of the surprising results in the study of convex sets is the way in which relatively weak local properties on certain sets imply that the whole set is convex. The earliest theorems of this sort were due to Tietze and Nakajima in 1928. We follow the approach used by Klee (1951a) in which he establishes the theorems of this section in the more general setting of a topological linear space. The techniques involved, as might be expected, are basically topological and depend on the connectedness of certain sets. (See Definition 1.24 and Exercise 1.19.) We begin with three definitions and two lemmas.

17.1. Definition. A subset S of E^n is **polygonally connected** if given any two points x and y in S there exist points

$$x_0 \equiv x, x_1, \ldots, x_{k-1}, x_k \equiv y,$$

such that

$$P \equiv \bigcup_{i=1}^{k} \overline{x_{i-1} x_i}$$

is contained in S. The set P is called a **polygonal path** from x to y.

LOCAL CONVEXITY

17.2. Definition. Let x be a point in E^n. A **neighborhood** of x is an open ball of positive radius centered at x.

17.3. Definition. A set S is **locally convex** if each point of S has a neighborhood whose intersection with S is convex.

17.4. Lemma. If S is a nonempty connected and locally convex set, then S is polygonally connected.

PROOF. Let $p \in S$ and let S_p denote the set of all points of S which can be joined to p by a polygonal path in S. We claim that S_p is both open and closed (as a subset of S). To see that S_p is open, let $q \in S_p$. Since S is locally convex, there exists a neighborhood N of q such that $N \cap S$ is convex. It follows that each point of $N \cap S$ can be joined to p by a polygonal path. Thus $N \cap S \subset S_p$ and S_p is an open subset of S.

To see that S_p is closed in S, let $z \in (\text{cl } S_p) \cap S$. Since S is locally convex there exists a neighborhood N of z such that $N \cap S$ is convex, and this neighborhood must also intersect S_p since $z \in \text{cl } S_p$. Thus there exists a point w in $(N \cap S) \cap S_p$. Since $N \cap S$ is convex, $\overline{wz} \subset S$. But w can be joined to p by a polygonal path in S, whence z can also. Thus $z \in S_p$ and S_p is closed in S.

Since S_p contains p, it is not empty. Since it is both open and closed in S and S is connected, it follows (Exercise 1.19) that S_p must be equal to S and so S is polygonally connected. ∎

17.5. Lemma. Let S be a closed, connected, locally convex set. If $\overline{xy} \subset S$ and $\overline{yz} \subset S$, then there exists a point q in $\overline{xy}$ with $q \neq y$, such that $\text{conv}\{q, y, z\} \subset S$.

PROOF. The only nontrivial case is that for which x, y, and z are not collinear. Let K be the set of all points p in $\overline{yz}$ such that $\text{conv}\{q, y, p\} \subset S$ for some q in $\overline{xy}$ with $q \neq y$. Since $y \in K$, K is not empty. We will show that K is both open and closed in $\overline{yz}$. Since $\overline{yz}$ is connected, this will imply that $K = \overline{yz}$, and the lemma will be proved.

To see that K is open in $\overline{yz}$, let $p \in K$. Then there exists a point $q \in \text{relint } \overline{xy}$ such that $\text{conv}\{q, y, p\} \subset S$. Since S is locally convex there exists a neighborhood N of p such that $N \cap S$ is convex. If $p \neq z$, choose $a \in N \cap \text{relint } \overline{pz}$ and $b \in N \cap \text{relint } \overline{pq}$. (See Figure 17.1.) Since a, b, and p are in the convex set $N \cap S$, we have $\text{conv}\{a, b, p\} \subset S$. The line through a and b intersects $\overline{xy}$ in a point c, which is in relint $\overline{qy}$. Since

$$\text{conv}\{c, y, a\} \subset \text{conv}\{q, y, p\} \cup \text{conv}\{a, b, p\},$$

we have $\overline{ya} \subset K$ and so K is open in $\overline{yz}$.

To see that K is closed in $\overline{yz}$, suppose $p \in \text{cl } K$, and let N be a neighborhood of p such that $N \cap S$ is convex. Since $p \in \text{cl } K$, there exists a point $a \in N \cap \text{relint } \overline{yp}$ and a point $q \in N \cap \text{relint } \overline{xy}$ such that $\text{conv}\{q, y, a\} \subset S$.

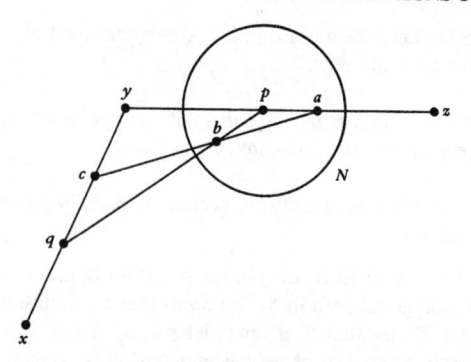

Figure 17.1.

(See Figure 17.2.) Choose $b \in N \cap \operatorname{relint} \overline{qa}$ and let c be the point where the line through b and p intersects relint $\overline{qy}$. As in the first part of the proof, we have $\operatorname{conv}\{a, b, p\} \subset S$ so that $\operatorname{conv}\{c, y, p\} \subset S$ and $p \in K$. Thus K is closed in $\overline{yz}$. ∎

17.6. Theorem. A closed connected set S is convex if and only if it is locally convex.

PROOF. Suppose that the closed connected set S is locally convex. Lemma 17.4 implies that S is polygonally connected, so in order to conclude that S is convex it suffices to show that $\overline{xy} \subset S$ and $\overline{yz} \subset S$ imply $\overline{xz} \subset S$.

To this end we suppose that $\overline{xy} \subset S$ and $\overline{yz} \subset S$. Let M be the set of all points r in $\overline{xy}$ such that $\operatorname{conv}\{r, y, z\} \subset S$. Since $y \in M$, M is not empty. The set M is closed, since S is closed. We will show that M is open in $\overline{xy}$ and thus that $M = \overline{xy}$ since $\overline{xy}$ is connected.

To see that M is open in $\overline{xy}$, let $r \in M$ with $r \neq x$. Then $\overline{xr} \subset S$ and $\overline{rz} \subset S$, so by Lemma 17.5 there exists a point q in $\overline{xr}$ with $q \neq r$ such that

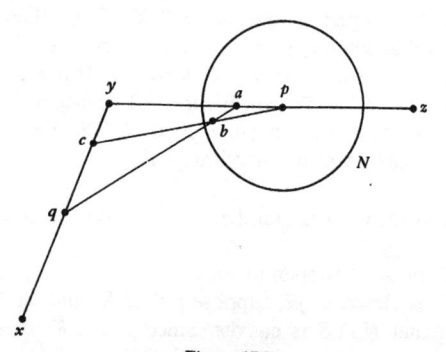

Figure 17.2.

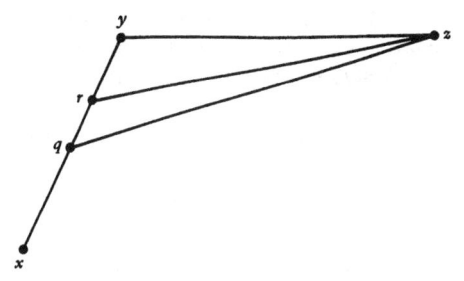

Figure 17.3.

conv$\{q, r, z\} \subset S$. (See Figure 17.3.) It follows that r is an interior point of M so that M is open in $\overline{xy}$. We thus conclude that $M = \overline{xy}$ and that $\overline{xz} \subset S$.

We have now shown that the local convexity of S implies that S is convex. The converse implication is immediate. ∎

EXERCISES

17.1. (a) Prove that every open set in E^n is locally convex.
  (b) Prove that every open connected set in E^n is polygonally connected.

17.2. A **crosscut** of a set S is defined to be a segment $\overline{xy}$ such that relint $\overline{xy} \subset$ int S and such that $x \in$ bd S, $y \in$ bd S. Prove the following: If an open set S has no crosscuts, then S is the complement of a convex set.

17.3. Let P be any property that is defined for all hyperplanes in E^n. Let x be a point such that each hyperplane through x has property P, and designate the set of all such points by S. Prove that each component of S is a convex set.

*17.4. Let S be a nonempty set and let F be a two-dimensional flat in E^n ($n \geq 3$). Then $S \cap F$ is called a **two-dimensional section** of S. Prove the following: Let S be a nonempty compact subset of E^n ($n \geq 3$). Then S is convex if and only if for each two-dimensional section K of S, both K and its complement $F \sim K$ are connected.

SECTION 18. LOCAL SUPPORT PROPERTIES

We now turn our attention to the problem of replacing the supporting hyperplanes of Theorem 5.4 by hyperplanes that support the set "locally." This concept is made precise in Definition 18.1 and the corresponding characterization is found in Theorem 18.8, which was first proved by Tietze in 1929. In following the approach of Valentine (1960) we shall find it convenient to introduce several related concepts of local support and derive two other characterizations of convex sets (Theorems 18.5 and 18.7).

18.1. Definition. Let S be a subset of E^n and let $x \in \text{bd } S$. Then S is **weakly supported** at x **locally** if there exists a neighborhood N of x and a linear functional f (not identically zero) such that the following holds:

$$\text{if } y \in N \sim \{x\} \quad \text{and} \quad f(y) > f(x), \quad \text{then } y \notin S.$$

For **strong local support** replace the inequality $f(y) > f(x)$ by $f(y) \geq f(x)$.

18.2. Definition. Let S be a subset of E^n. The point $x \in \text{bd } S$ is a point of **strong local concavity** if there exists a neighborhood N of x and a linear functional f (not identically zero) such that the following holds:

$$\text{if } y \in N \sim \{x\} \quad \text{and} \quad f(y) \leq f(x), \quad \text{then } y \in S.$$

For **weak local concavity** replace the inequality $f(y) \leq f(x)$ by $f(y) < f(x)$.

18.3. Example. Figure 18.1 illustrates each of the above ideas. The set S is weakly supported locally at x_1 and strongly supported locally at x_2, x_3 and x_4. The point x_2 is a point of weak local concavity, and the points x_1 and x_5 are points of strong local concavity. Of course, if S is strongly supported locally at a point y, then it is also weakly supported locally at y. Furthermore, if y is a point of strong local concavity, then it is also a point of weak local concavity.

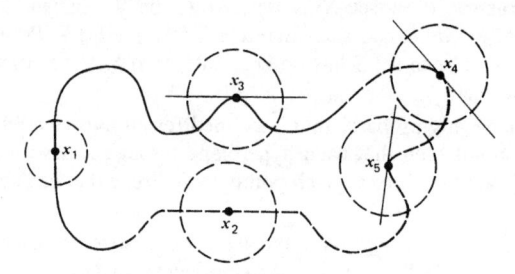

Figure 18.1.

Our first characterization theorem equates convexity in E^2 with the absence of boundary points of strong local concavity. Its proof is built on the following lemma, which involves the existence of exposed points. (See Exercise 5.9.)

18.4. Lemma. Let S be a closed convex subset of E^2. Each compact connected portion of the boundary of S that is not contained in a line segment contains an exposed point of S. If S is a line segment, it has two exposed points.

PROOF. Suppose first that S is compact. For each line of support H to S there exists exactly one line R through the origin θ that is parallel to H. Let $H \cap \text{bd } S = P$. If the boundary of S contained no exposed points of S, then P

LOCAL SUPPORT PROPERTIES

would have to be a nondegenerate line segment. But since S is compact, its boundary can contain only a countable number of nonoverlapping segments. Since there are uncountably many lines R through the origin, at least one of the corresponding lines of support to S must meet S in a single point. Hence the boundary of S contains at least one exposed point.

If K is a compact connected portion of the boundary of S that is not contained in a line segment, then an argument similar to the above will imply the existence of an exposed point of S lying in K.

The final assertion of the lemma is trivial. ∎

18.5. Theorem. Let S be an open connected subset of E^2. Then S is convex if and only if S has no point of strong local concavity in its boundary.

PROOF. Suppose first that S is not convex. We must show that S has a point of strong local concavity in its boundary. By Exercise 17.1(b), S is polygonally connected. Thus there exist noncollinear points x, y, z in S such that $\overline{xz} \subset S$, $\overline{zy} \subset S$, but $\overline{xy} \not\subset S$. Since S is open, there exists a point w on the line through x and z such that $x \in \text{relint}\,\overline{wz}$ and $\overline{wz} \subset S$. (See Figure 18.2.) Define

$$S^* \equiv \text{conv}\big[(\text{bd}\,S) \cap (\text{conv}\{w, y, z\})\big].$$

By Lemma 18.4, there exists a point p which is an exposed point of S^*. Since $\overline{xy} \cap \text{bd}\,S \neq \varnothing$, p can be chosen so that $p \notin \overline{wy}$. It follows readily from the definition of an exposed point (Exercise 18.1) that $p \in \text{bd}\,S$. We claim that p is a point of strong local concavity of S.

Since $p \in \text{int}\,\text{conv}\{w, y, z\}$, there exists a neighborhood N of p such that $N \subset \text{conv}\{w, y, z\}$. Since p is an exposed point of S^*, there exists a hyperplane (line) $H \equiv [f: \alpha]$ such that $f(p) = \alpha$ and $f(q) > \alpha$ for all $q \in S^* \sim \{p\}$. It follows that for $q \in N \sim \{p\}$ with $f(q) \leq f(p)$ we must have $y \in S$. Thus p is a point of strong local concavity of S.

We have now shown that if S is not convex then it must have a point of strong local concavity in its boundary. The converse follows directly, and is left to the reader (Exercise 18.2). ∎

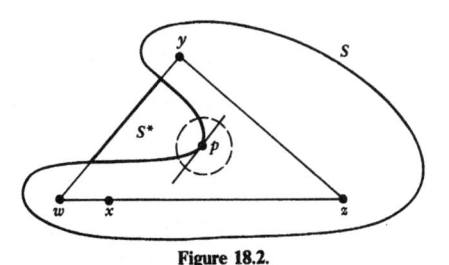

Figure 18.2.

It should be pointed out that Theorem 18.5 does not hold for E^n with $n \geq 3$. For example, let S be the union of two open intersecting spheres. Theorem 18.5 will be useful, however, in higher dimensions when we can restrict the problem to a two-dimensional flat. This is utilized in Theorem 18.7, which is preceded by a definition.

18.6. Definition. Let S be a subset of E^n and let $x \in$ bd S. Then x is a **point of mild convexity** of S if no nondegenerate line segment $\overline{uv}$ exists having x as midpoint and having $\overline{uv} \sim \{x\} \subset$ int S.

18.7. Theorem. Let S be an open connected subset of E^n. Then S is convex if and only if each boundary point of S is a point of mild convexity of S.

PROOF. Suppose first that each boundary point of S is a point of mild convexity of S. We again use the fact (Exercise 17.1) that S is polygonally connected, and show that $\overline{xz} \subset S$, $\overline{zy} \subset S$ implies $\overline{xy} \subset S$. To this end, let H_2 be a two-dimensional flat containing x, y and z, and let K be the component of $S \cap H_2$ that contains $\{x, y, z\}$. Now K is an open set relative to H_2, and each of the boundary points of K is a point of mild convexity of K. Since in H_2 a point of mild convexity is not a point of strong local concavity (Exercise 18.4), Theorem 18.5 implies that K is convex. Thus $\overline{xy} \subset K \subset S$ and S is convex.
The converse is left to the reader (Exercise 18.5). ∎

We are now in a position to prove Tietze's theorem, which characterizes convexity in terms of the weak local support property.

18.8. Theorem (Tietze). Let S be an open connected subset of E^n. Then S is convex if and only if S is weakly supported locally at each of its boundary points.

PROOF. The sufficiency part of this theorem will follow from Theorem 18.7 if we can show that each boundary point of S at which S is weakly supported locally is a point of mild convexity of S. To this end, suppose S is weakly supported locally at x and let N and f be the neighborhood and linear functional (respectively) of Definition 18.1. If x is not a point of mild convexity of S, then there exists a nondegenerate segment $\overline{uv}$ having midpoint x with

$$\overline{uv} \sim \{x\} \subset [N \cap \text{int } S].$$

(See Figure 18.3.) Since either $f(u) > f(x)$ or $f(v) > f(x)$ violates our hypotheses and $x = (u + v)/2$, we must have

$$f(u) = f(x) = f(v).$$

Choose a vector w so that $z \equiv v + \alpha w$ satisfies $f(z) > f(x)$. Since $v \in$ int S, we have $z \in$ int S for $|\alpha|$ sufficiently small. Hence there exists a point z in

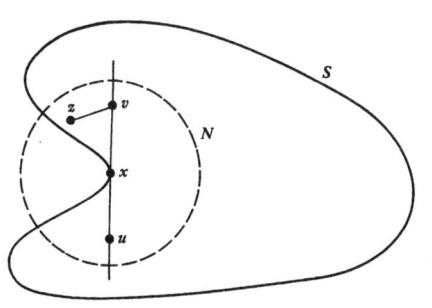

Figure 18.3.

(int S) $\cap$ ($N \sim \{x\}$) such that $f(z) > f(x)$, a contradiction. We conclude that x must be a point of mild convexity, and the sufficiency half of the theorem is proved.

The necessity half of the theorem is left to the reader (Exercise 18.6). ∎

EXERCISES

18.1. Let K be a compact subset of E^n and let x be an exposed point of conv K. Prove that $x \in K$.

18.2. Let S be an open subset of E^2. Prove the converse part of Theorem 18.5: If there exists a point of strong local concavity in the boundary of S, then S is not convex.

18.3. Which of the points in Figure 18.1 are points of mild convexity of S?

18.4. Let S be an open subset of E^2 and let $x \in$ bd S. Prove that if x is a point of mild convexity of S then x is not a point of strong local concavity of S.

18.5. Let S be an open connected subset of E^n. Prove the converse part of theorem 18.7: If S is convex, then each boundary point of S is a point of mild convexity of S.

18.6. Let S be an open connected subset of E^n. Prove the necessity half of Theorem 18.8: If S is convex, then S is weakly supported locally at each of its boundary points.

SECTION 19. NEAREST-POINT PROPERTIES

Suppose S is a nonempty closed convex subset of E^n. It is easy to see that for any point x in E^n, there is a unique point of S that is nearest to x. Indeed, since S is closed, there exists at least one such nearest point. If there were two such points, say y and z, then conv$\{x, y, z\}$ is an isosceles triangle. It follows that the point $w \equiv (y + z)/2$ is closer to x than y and z. (See Figure 19.1.) But since S is convex, $w \in S$ and this contradicts y and z as being nearest points of S to x.

```
         •y
    ╱─────────•
x•╱           w
   ╲─────────•
         •z   Figure 19.1.
```

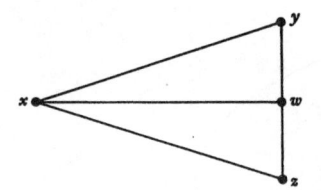

It is quite remarkable that the converse to the above is also true. This was first proved by T. S. Motzkin in 1935. We begin our investigation of this topic with a definition.

19.1. Definition. Let S be a nonempty closed subset of E^n. Then S has the **nearest-point property** if for each $x \in \mathsf{E}^n$ there exists a unique point y (depending on x) such that y is the nearest point of S to x. That is, for each x there exists exactly one point y that satisfies

$$d(y, x) = \inf_{s \in S} d(s, x).$$

19.2. Theorem. Let S be a nonempty closed subset of E^n. Then S is convex if and only if S has the nearest-point property.

PROOF. Suppose that S has the nearest-point property but that S is not convex. Then there exist points x and y in bd S such that $S \cap \text{relint}\,\overline{xy} = \varnothing$. Since S is closed, there exists a closed sphere K centered at the midpoint z of $\overline{xy}$ and having positive radius such that $K \cap S = \varnothing$. Let $\mathcal{F}$ denote the collection of all closed spheres C satisfying:

$$K \subset C \quad \text{and} \quad S \cap \text{int}\,C = \varnothing.$$

We now claim that $\mathcal{F}$ is a compact family. Indeed, since $K \subset C$, $x \notin \text{int}\,C$ and $y \notin \text{int}\,C$ for $C \in \mathcal{F}$, the family $\mathcal{F}$ is uniformly bounded. Blaschke's Theorem 15.2 thus implies that each sequence in $\mathcal{F}$ contains a convergent subsequence, and since the limit of such a convergent subsequence will contain K and have an interior disjoint from S, it follows that $\mathcal{F}$ is a compact family.

Therefore, we can apply Theorem 16.1 to obtain a sphere C^* in $\mathcal{F}$ having a maximum radius. Since S has the nearest-point property and C^* is maximal, C^* must intersect S in exactly one point, say q. We now translate C^* away from q toward a point "p" as follows: if $K \cap \text{bd}\,C^* = \varnothing$, let $p = z$. If $K \cap \text{bd}\,C^* \neq \varnothing$, then since K and C^* are both spheres, this intersection must consist of a single point, and we call this point p. (See Figure 19.2.) In either case, $p \neq q$ and relint $\overline{pq} \subset \text{int}\,C^*$. Since S is closed, there exists a positive number λ such that the translate

$$C^* + \lambda(p - q)$$

Figure 19.2.

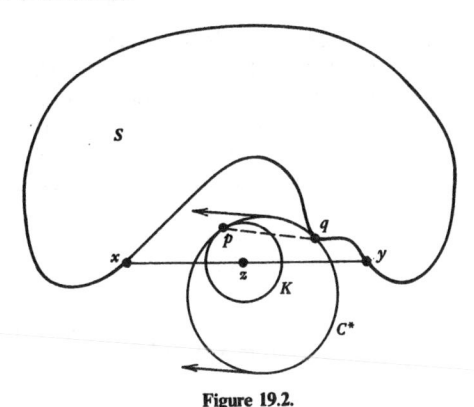

contains K and is disjoint from S. It follows that there exists a sphere C_1 in $\mathcal{F}$ concentric to $C^* + \lambda(p - q)$ and properly containing it. This, however, contradicts the maximality of the radius of C^*. Hence our assumption that S is not convex leads to a contradiction, and we conclude that if S has the nearest-point property then S must be convex. The converse is proved in the comments at the beginning of this section. ∎

It should be pointed out that Theorem 19.2 is dependent on the fact that in Euclidean space the unit ball is "smooth" and "round." To be precise, this means that at each point of its boundary there is a unique hyperplane of support, and the boundary contains no straight-line segments. This implies that when the two unequal spheres K and C^* have a boundary point in common (and $K \subset C^*$), then this common boundary point is unique. To see that this is not always the case for non-Euclidean spaces, we consider the following example.

19.3. Example. On the set of all ordered pairs of real numbers R^2, define a new distance function d_2 by

$$d_2[(x_1, y_1), (x_2, y_2)] \equiv \max\{|x_1 - x_2|, |y_1 - y_2|\}$$

for (x_1, y_1) and (x_2, y_2) in R^2. Using this distance function it is possible to derive a topology for R^2 in the same way we derived the usual Euclidean topology in Section 1. We will denote this new space by D^2. It is easy to see (Exercise 19.2) that the D^2 topology is "equivalent" to the Euclidean topology in the sense that a set which is open in one will also be open in the other.

While the new topology is essentially the same as Euclidean, and the linear structures are identical, the geometrical properties of D^2 are quite different from E^2. For example, the closed unit "ball" is no longer round—it is the square with vertices at $(1,1)$, $(1,-1)$, $(-1,-1)$, $(-1,1)$. Because of this difference, Theorem 19.2 does not hold in D^2. Indeed, the closed unit ball is a convex set that does not have the nearest-point property: All the points along the line segment from $(1,1)$ to $(1,-1)$ are at a d_2-distance of 2 from the point $(3,0)$. It is also possible to construct a nonconvex set that *has* the nearest-point property (see Exercise 19.3).

We conclude this section with a theorem based on the idea of a unique farthest point. It was first proved by Motzkin, Straus, and Valentine in 1953.

19.4. Theorem. Let S be a closed subset of E^n. Then for each $x \in E^n$ there exists a unique farthest point of S from x if and only if S consists of a single point.

PROOF. Suppose first that for each x in E^n there exists a unique farthest point of S from x. Then in particular, there exists a farthest point of S from the origin θ, so S must be bounded. Let K be a closed sphere about θ containing $2S$ and let $\mathcal{F}$ be the family of all closed spheres C such that $S \subset C \subset K$. As in the proof of Theorem 19.2, we find that $\mathcal{F}$ is a compact family. Theorem 15.6 then implies the existence of a sphere C^* in $\mathcal{F}$ having a *minimum* radius r. If $r = 0$, then C^* is a point and hence S is a point.

Thus we suppose that $r > 0$, and without loss of generality assume that C^* is centered at θ. Since S has a unique farthest point from θ, the minimality of C^* implies that S intersects C^* in a unique point, say p. But then there exists a positive number λ such that the translate

$$C^* + \lambda p$$

contains S in its interior and is contained in K (see Figure 19.3). Since S is closed, it follows that there exists a sphere in $\mathcal{F}$ of radius less than r, a contradiction. Thus we must conclude that $r = 0$ and that S consists of a single point.

The converse implication is immediate. ∎

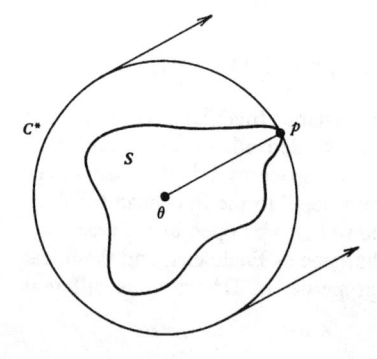

Figure 19.3.

EXERCISES

19.1. Let d_2 be the distance function defined in Example 19.3. Verify that d_2 has all the properties listed in Theorem 1.6.

19.2. Let D^2 be the metric space defined in Example 19.3. Verify that a set is open in D^2 is and only if it is open in E^2.

19.3. Let D^2 be the metric space defined in Example 19.3. Find a nonconvex set in D^2 that *has* the nearest-point property.

19.4. Define a distance function d^* on R^2 by

$$d^*[(x_1, y_1), (x_2, y_2)] = |x_1 - x_2| + |y_1 - y_2|$$

for $(x_1, y_1), (x_2, y_2)$ in R^2. Let D^* denote the resulting topological space.
(a) Verify that d^* satisfies all the properties listed in Theorem 1.6.
(b) Prove that the topology derived from d^* is equivalent to the Euclidean topology.
(c) Find a convex set in D^* that does not have the nearest-point property.
(d) Find a nonconvex set in D^* that has the nearest-point property.

19.5. Let S be a nonempty compact convex subset of E^n. For each $x \in E^n$, Theorem 19.2 implies that there exists a unique point x' in S that is nearest to S. Let Ψ be the mapping defined by $\Psi(x) = x'$.
(a) Let $x \in E^n \sim S$ so that $\Psi(x) \neq x$, and let R be the ray (half-line) starting at $\Psi(x)$ that passes through x. Prove that if $y \in R$, then $\Psi(y) = \Psi(x)$.
(b) Let x and R be chosen as in part (a). Prove that the hyperplane H that passes through $\Psi(x)$ and that is orthogonal to R supports S.
(c) Use parts (a) and (b) to prove the following: Every compact convex subset of E^n is the intersection of all its supporting half-spaces. (A supporting half-space to S is a half-space containing S and bounded by a supporting hyperplane to S.)
(d) Prove that the mapping Ψ does not increase length. That is, for all $x, y \in E^n$ we have $\|x - y\| \geq \|\Psi(x) - \Psi(y)\|$.
(e) Prove that Ψ is a continuous mapping.
(f) Let K be the surface of a closed sphere of positive radius that contains S. Prove that Ψ maps K onto bd S.
(g) Use part (f) to give an alternate proof of Theorem 5.2: If x is a boundary point of S, then there exists at least one hyperplane supporting S at x.

†19.6. Generalize Theorem 19.2 to a complete infinite-dimensional inner-product space (i.e., a Hilbert space).

8
POLYTOPES

Section 2 defined a (convex) polytope to be the convex hull of a finite set of points. This is the natural generalization of a polygon in the plane and a polyhedron in three-space. These lower dimensional figures have been studied since the time of the ancient Greeks more than two thousand years ago. During the last fifty years there has been a renewed interest in polytopes of all dimensions because of their importance in linear programming and game theory.

In this chapter we will present only a brief introduction to the theory of polytopes. The interested reader is referred to the substantial books by Grünbaum (1967) and Coxeter (1963) and the smaller paperback by McMullen and Shephard (1971) for a more expansive treatment of the subject.

SECTION 20. THE FACES OF A POLYTOPE

In this section we will introduce the basic terminology for working with polytopes and derive some of the fundamental theorems.

20.1. Definition. Let S be a compact convex subset of E^n. A subset F of S is called a **face** of S if either $F = \emptyset$ or $F = S$, or if there exists a supporting hyperplane H of S such that $F = S \cap H$. The faces S and $\emptyset$ are called the **improper** faces of S. All other faces are called **proper**. If the dimension of F is k, then F is called a **k-face** of S. If P is a polytope of dimension k, then P is called a **k-polytope**.

It is customary to refer to the 0-faces of S as **vertices** (or exposed points) and the set of all vertices of S will be denoted by vert S. The 1-faces of S are called **edges** and the $(n-1)$-faces are called **facets** of S.

One might expect the vertices of a polytope $P \equiv \text{conv}\{x_1, \ldots, x_m\}$ to be the extreme points of P (cf. Definition 5.5) and to coincide with the points $x_1, \ldots, x_m$. This will be true if the set $\{x_1, \ldots, x_m\}$ is minimal in

THE FACES OF A POLYTOPE

the following sense:

20.3. Definition. The set $\{x_1, \ldots, x_k\}$ is a **minimal representation** of the polytope P if $P = \text{conv}\{x_1, \ldots, x_k\}$ and for each $i = 1, \ldots, k$, $x_i \notin \text{conv} \cup_{j \neq i} x_j$.

It is clear that every polytope has a minimal representation and that this representation is unique. Indeed, if P is expressed as the convex hull of the points $x_1, \ldots, x_m$ and some point x_j is a convex combination of the other points, then x_j may be deleted from the set of points without changing the convex hull. This process may be repeated until the minimal representation is left.

The following theorem verifies our earlier remarks about vertices and shows that our definition of vertex in 20.1 for compact convex sets is consistent with our earlier definition 2.24 for simplices.

20.4. Theorem. Suppose $M \equiv \{x_1, \ldots, x_k\}$ is the minimal representation of the polytope P. Then the following three statements are equivalent:
  (a) $x \in M$.
  (b) x is a vertex of P.
  (c) x is an extreme point of P.

PROOF. (a) $\Rightarrow$ (b) Suppose $x \in M$ and let $Q = \text{conv}(M \sim \{x\})$. It follows from the definition of M that $x \notin Q$, and since Q is compact, Theorem 4.12 implies the existence of a hyperplane H' that strictly separates $\{x\}$ and Q. Let H be the hyperplane through x parallel to H'. (See Figure 20.1.) Then Q lies in one of the closed half-spaces H^+ bounded by H and so $P \subset H^+$. Thus H supports P at x. Furthermore, x is the only point of P that can lie on H, so $H \cap P = \{x\}$ and x is a vertex of P.

(b) $\Rightarrow$ (c) A vertex of a compact convex set is always an extreme point, even if the set is not a polytope. The proof of this is left for Exercise 20.1.

(c) $\Rightarrow$ (a) It is clear that any extreme point of P must be a member of M. ∎

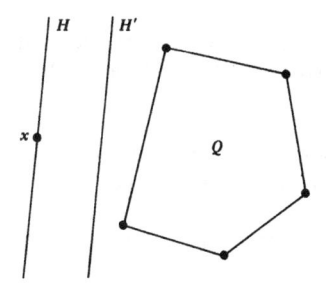

Figure 20.1.

20.5. Theorem. Let P be a polytope in $\mathbf{E}^n$. Then each proper face of P is itself a polytope, and there are only a finite number of distinct faces.

PROOF. Let $\{x_1, \ldots, x_k\}$ be the minimal representation of P and suppose that F is a proper face of P whose supporting hyperplane is given by $H \equiv [f: \alpha]$ for some linear functional f. Without loss of generality we may suppose that $\{x_1, \ldots, x_r\} \subset H$ and that $f(x_i) > \alpha$ for $i = r+1, \ldots, k$. That is,

$$f(x_i) = \alpha \qquad (i = 1, \ldots, r)$$

and

$$f(x_i) = \alpha + \varepsilon_i \quad \text{for some} \quad \varepsilon_i > 0 \qquad (i = r+1, \ldots, k)$$

Now let

$$x \equiv \lambda_1 x_1 + \cdots + \lambda_k x_k, \qquad \lambda_i \geq 0, \qquad \sum_{i=1}^{k} \lambda_i = 1,$$

be an arbitrary point of P. Then

$$f(x) = \lambda_1 \alpha + \cdots + \lambda_r \alpha + \lambda_{r+1}(\alpha + \varepsilon_{r+1}) + \cdots + \lambda_k(\alpha + \varepsilon_k)$$

$$= \alpha \sum_{i=1}^{k} \lambda_i + \sum_{i=r+1}^{k} \lambda_i \varepsilon_i$$

$$= \alpha + \sum_{i=r+1}^{k} \lambda_i \varepsilon_i.$$

The point x lies in H iff $f(x) = \alpha$. That is, if and only if $\sum_{i=r+1}^{k} \lambda_i \varepsilon_i = 0$. Since each ε_i is positive and each λ_i is nonnegative, this will happen iff $\lambda_{r+1} = \cdots = \lambda_k = 0$. But then x is a convex combination of $x_1, \ldots, x_r$. Hence we conclude that

$$H \cap P = \text{conv}\{x_1, \ldots, x_r\},$$

and so $H \cap P$ is a polytope.

Finally, since the set $\{x_1, \ldots, x_k\}$ has only a finite number of subsets, and since each face of P corresponds to one of these subsets, P can have only a finite number of distinct faces. ∎

While Theorem 20.5 helps to describe the faces of a polytope, the following theorem applies equally well to the faces of any compact convex set.

20.6. Theorem. Let $\{F_1, \ldots, F_m\}$ be a family of faces of a compact convex set S. Then $\bigcap_{i=1}^{m} F_i$ is also a face of S.

THE FACES OF A POLYTOPE

PROOF. Let $F \equiv \bigcap_{i=1}^{m} F_i$. If $F = \varnothing$, then the result is true by definition. Thus we take $F \neq \varnothing$, and we may assume without loss of generality that the origin θ belongs to F and that each F_i is a proper face of S. Thus each F_i is contained in a supporting hyperplane H_i and since $\theta \in F_i$, there exist linear functionals f_i such that

$$H_i = [f_i : 0] \quad \text{and} \quad f_i(S) \geq 0 \quad \text{for all } i.$$

We now define a new linear functional f by

$$f \equiv \sum_{i=1}^{m} f_i,$$

and let $H \equiv [f : 0]$. We claim that $F = H \cap S$.

Indeed, if $x \in F$, then $f_i(x) = 0$ for all i and so $f(x) = 0$. Thus $x \in H \cap S$ and $F \subset H \cap S$. On the other hand, if $x \in S \sim F$, then $f_i(x) > 0$ for at least one i, and so $f(x) > 0$. Thus $x \notin H \cap S$ and we conclude that $F = H \cap S$.

Finally, since $\theta \in H \cap S$ and $f(S) \geq 0$, H is a supporting hyperplane to S and it follows that F is a face of S. ∎

In Section 4 (Exercise 4.2) we observed that a closed convex set is the intersection of a family of closed half-spaces. For polytopes, a finite family suffices, as proved in Theorem 20.8. The converse of this is proved in Theorem 20.9, and together they completely characterize polytopes. The following definition is useful.

20.7. Definition. A **polyhedral set** is the intersection of a finite number of closed half-spaces.

20.8. Theorem. Every polytope P in E^n is a bounded polyhedral set.

PROOF. Let $\{x_1, \ldots, x_k\}$ be the minimal representation of P and assume, without loss of generality, that P is n-dimensional. Let $F_1, \ldots, F_m$ denote the facets of P, and for each F_i let H_i and H_i^+ denote the associated hyperplane and closed half-space (respectively) such that $F_i = H_i \cap P$ and $P \subset H_i^+$. We claim that

$$P = H_1^+ \cap \cdots \cap H_m^+.$$

In the first place, suppose that there exists a point x in $H_1^+ \cap \cdots \cap H_m^+$ with $x \notin P$. We define

$$D \equiv \bigcup_B \text{aff}(\{x\} \cup B),$$

where B is any subset of $n - 1$ or fewer points of $\{x_1, \ldots, x_k\}$. Then D consists of a finite union of flats of dimension at most $n - 1$. Since $\dim P = n$, this

Figure 20.2.

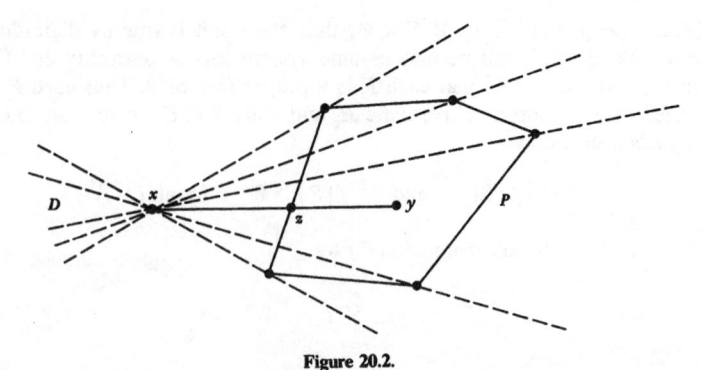

implies that int $P \not\subset D$ and so there exists a point y in (int P) $\sim D$. (See Figure 20.2.)

Since $y \in$ int P and $x \notin P$, there exists a point z in relint $\overline{xy}$ such that $z \in$ bd P (cf. Exercise 2.11). We claim that z belongs to a facet of P and to no face of lower dimension. Indeed, if z belonged to a j-face of P with $0 \leq j \leq n - 2$, then by Caratheodory's Theorem 2.23, $z \in$ conv B_0, where B_0 is some subset of $n - 1$ or fewer points of $\{x_1, \ldots, x_m\}$. This implies that $z \in D$, and since $x \in D$, we have $y \in D$, a contradiction.

Thus we conclude that z belongs to a facet of P, say F_j. Then $z \in H_j$, and since $y \in$ int $P \subset H_j^+$, x cannot lie in H_j^+. This contradicts our initial supposition that $x \in H_1^+ \cap \cdots \cap H_m^+$ and so we must have

$$H_1^+ \cap \cdots \cap H_m^+ \subset P.$$

The reverse inclusion follows easily since $P \subset H_i^+$ for each i. Thus P is a polyhedral set, and it clearly is bounded. ∎

20.9. Theorem. Every bounded polyhedral set P in E^n is a polytope.

PROOF. Since every bounded polyhedral set is compact, it suffices by Theorem 5.6 to show that P has a finite number of extreme points. Our proof is by induction on the dimension n of the space.

When $n = 1$, P is either a single point or a closed line segment and thus has a finite number of extreme points.

Assume inductively that any polyhedral set in an $(n - 1)$-dimensional space has a finite number of extreme points. Let $H_1, \ldots, H_m$ denote the hyperplanes that bound the half-spaces of which P is the intersection. If x is an extreme point of P, then $x \in$ bd P, and so $x \in H_i$ for some $i = 1, \ldots, m$. But an extreme point of P in H_i is an extreme point of $P \cap H_i$ (see Exercise 5.5), and there are only a finite number of such points by the induction hypothesis. Since the number of hyperplanes is finite, P can have at most a finite number of extreme points. ∎

THE FACES OF A POLYTOPE

If a polyhedral set is bounded, then we have just seen that it is a polytope. But even if it is not bounded, it still has some special properties. One of these is included in the following theorem, which will be of particular importance in our development of linear programming (see Section 26).

20.10. Theorem. Let P be a polyhedral subset of E^n that contains no lines, and suppose the linear functional $f\colon \mathsf{E}^n \to \mathsf{R}$ is bounded above on P. [That is, $\sup_{x \in P} f(x) < \infty$.] Then there exists an extreme point $\bar{x}$ of P such that

$$f(\bar{x}) = \sup_{x \in P} f(x).$$

PROOF. We may assume without loss of generality that $\dim P = n$. The proof is by induction on n. If $n = 0$, the result is trivial. Suppose now that the theorem holds for spaces of dimension less than n. Since f is linear, it is not difficult to show (Exercise 3.12) that

$$\alpha \equiv \sup_{x \in P} f(x) = \sup_{x \in \mathrm{bd}\, P} f(x).$$

Since P is polyhedral, its boundary can be written as $(P \cap H_1) \cup \cdots \cup (P \cap H_m)$, where $H_1, \ldots, H_m$ are the hyperplanes that bound the half-spaces whose intersection is P. Thus

$$\alpha = \sup_{x \in P \cap H_i} f(x)$$

for some i. We claim that $P \cap H_i$ is also polyhedral. Indeed, if $n = 1$, then $P \cap H_i$ is a single point. If $n > 1$, then since P contains no lines, $P \cap H_i \neq H_i$. In either case, $P \cap H_i$ is polyhedral, and it certainly contains no lines. Since $\dim(P \cap H_i) < n$, our induction hypothesis implies that $\alpha = f(\bar{x})$ for some extreme point $\bar{x}$ in $P \cap H_i$. But any extreme point of $P \cap H_i$ is an extreme point of P (Exercise 5.5), so we are done. ∎

We conclude Section 20 with two theorems that relate faces and subsets. They will be particularly useful in Section 21 when we count the number of faces of a given dimension for several special polytopes.

20.11. Theorem. Suppose S_1 and S_2 are compact convex sets such that $S_2 \subset S_1$. If F is a face of S_1, then $F \cap S_2$ is a face of S_2.

PROOF. If F is an improper face of S_1, then the result follows trivially. Otherwise, let H be a supporting hyperplane of S_1 with $H \cap S_1 = F$. Since $S_2 \subset S_1$, it follows that $F \cap S_2 = H \cap S_2$. Now either $H \cap S_2$ is empty or H supports S_2. In either case $F \cap S_2$ is a face of S_2. ∎

20.12. Theorem. Suppose F_1 is a face of the polytope P and F_2 is a face of F_1. Then F_2 is a face of P.

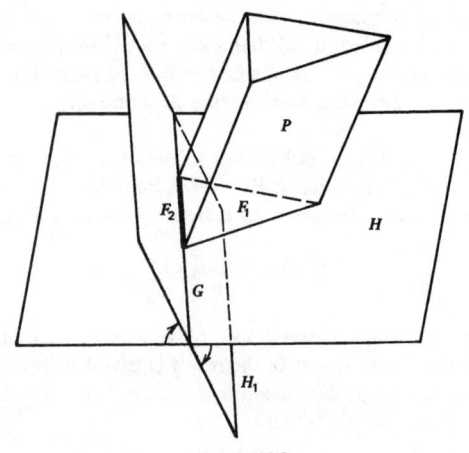

Figure 20.3.

PROOF. The idea of this proof is fairly simple, although writing out the details can be a bit messy. We will outline the idea here and leave the interested reader to consult Grünbaum (1967) for the details.

Let H be a supporting hyperplane to P such that $H \cap P = F_1$ and let H' be a supporting hyperplane to F_1 such that $H' \cap F_1 = F_2$. Since H and H' both contain F_2 but only H contains F_1, it follows that $G \equiv H \cap H'$ is an $(n-2)$-dimensional flat. (See Figure 20.3.) Since P has only finitely many vertices not in F_2, a sufficiently small rotation of H about G in a direction "away" from these vertices will produce a new hyperplane H_1 which still supports P and such that $H_1 \cap P = F_2$. ∎

EXERCISES

20.1. (a) Prove that every vertex of a compact convex set S is an extreme point of S.
  (b) Find an example in E^2 to show that the converse of part (a) does not hold when S is not a polytope.

20.2. Suppose P is a polytope and L is a flat in E^n. Prove that $L \cap P$ is a polytope.

20.3. Suppose F_1 is a compact convex subset of a compact convex set S. If $F_2 \subset F_1$ and F_2 is a face of S, prove that F_2 is a face of F_1.

20.4. Find an example to show that Theorem 20.12 only holds for polytopes. That is, find a compact convex set S having a face F_1 where F_1 has a face F_2 such that F_2 is not a face of S.

20.5. Let $P_1, \ldots, P_k$ be a finite number of polytopes. Prove that each of the following sets is also a polytope.
  (a) $\text{conv}(P_1 \cup \cdots \cup P_k)$.

SPECIAL TYPES OF POLYTOPES AND EULER'S FORMULA

(b) $\bigcap_{i=1}^{k} P_i$.
(c) The vector sum $P_1 + \cdots + P_k$.

20.6. Suppose that P is a polytope and that $\delta > 0$. Prove that δP is also a polytope.

20.7. Show that both of the restrictions on P in Theorem 20.10 are necessary. Specifically, do the following:
(a) Find a polyhedral set S and a linear functional f that is bounded above on S, but such that f does not attain its upper bound at an extreme point of S.
(b) Do the same thing when S is a closed convex set that contains no lines.

20.8. Let $a_1 = (0,0,0)$, $a_2 = (3,0,0)$, $a_3 = (0,3,0)$, and $a_4 = (1,1,3)$, and let $P = \text{conv}\{a_1, a_2, a_3, a_4\}$.
(a) Let $f_{ijk}(x, y, z)$ denote a linear functional such that $[f_{ijk}: 0]$ intersects P in the face which contains a_i, a_j, and a_k, and such that $f_{ijk}(P) \geq 0$. Find f_{124}, f_{134}, f_{123}.
(b) Verify that $f_{14} \equiv f_{124} + f_{134}$ is a linear functional such that $[f_{14}: 0]$ supports P along the face (edge) $\text{conv}\{a_1, a_4\}$ (cf. Theorem 20.6).

20.9. Two polytopes P and Q are said to be **combinatorially equivalent** (or **isomorphic** or of the same **combinatorial type**) if there exists a one-to-one correspondence Ψ between the set of faces of P and the set of faces of Q such that Ψ is inclusion-preserving; that is, for any two faces F_1 and F_2 of P, $F_1 \subset F_2$ iff $\Psi(F_1) \subset \Psi(F_2)$.
(a) Prove that all polytopes in E^1 are combinatorially equivalent to each other.
(b) Characterize the different combinatorial types in E^2.
†(c) Characterize the different combinatorial types in E^n for $n \geq 3$.

SECTION 21. SPECIAL TYPES OF POLYTOPES AND EULER'S FORMULA

In this section we look carefully at some particular examples of polytopes that arise as natural generalizations of familiar polygons and polyhedra. This should help to strengthen the reader's intuition for geometric relationships in higher dimensions. Throughout this section we let $f_k(P)$ denote the number of k-faces of an n-polytope P in E^n ($0 \leq k \leq n-1$). As we compute these numbers for each polytope, we will observe that they satisfy Euler's remarkable formula

$$\sum_{k=0}^{n-1} (-1)^k f_k(P) = 1 + (-1)^{n-1},$$

and finally we shall prove this result for $n = 3$.

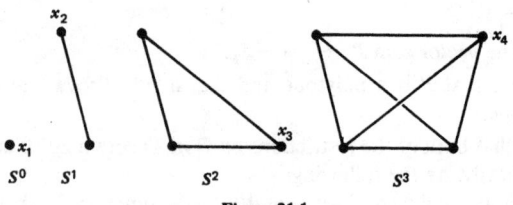

Figure 21.1.

Simplices and Pyramids

The simplest type of polytope is the simplex that was introduced in Section 2 (see Definition 2.24). A k-simplex (k-dimensional simplex) can be constructed in the following way:

0-simplex S^0:   a single point $\{x_1\}$
1-simplex S^1: $\text{conv}(S^0 \cup \{x_2\})$ where $x_2 \notin \text{aff } S^0$
2-simplex S^2: $\text{conv}(S^1 \cup \{x_3\})$ where $x_3 \notin \text{aff } S^1$
$\vdots$
k-simplex S^k: $\text{conv}(S^{k-1} \cup \{x_{k+1}\})$ where $x_{k+1} \notin \text{aff } S^{k-1}$

The simplex S^1 is a line segment. By joining a point x_3 which is not in the line containing S^1, we obtain the triangle S^2.* (See Figure 21.1.) By joining point x_4 (not in the plane of S^2) we obtain the tetrahedron S^3.

If we join a point x_5 (not in the three-space of S^3) to S^3, we obtain the pentatope S^4. At this stage it becomes more difficult to draw an accurate picture, but the representation in Figure 21.2a is suggestive: S^4 has five vertices, and any four of the vertices determine a facet in the shape of a tetrahedron. For example, in Figure 21.2b we have emphasized the facet with vertices x_1, x_2, x_4, and x_5. Altogether there are five such facets. There are also ten triangular faces and ten edges. (Count them!)

In general, every face of S^n is the convex hull of some subset of the vertices, and since every subset of an affinely independent set is affinely independent, it is itself a simplex S^k for $k \leq n$. Since any set M consisting of n vertices of S^n affinely spans a supporting hyperplane of S^n, $\text{conv } M$ is an $(n-1)$-simplex that is a facet of S^n. Since every face of $\text{conv } M$ is a face of S^n (Theorem 20.11), we obtain the following theorem by induction.

21.1. Theorem. Every k-face ($0 \leq k \leq n - 1$) of an n-simplex S^n is a k-simplex, and every $k + 1$ vertices of S^n are the vertices of a k-face of S^n.

---

*We have spoken intuitively of "joining" a point x_3 to a convex set S^1. Technically, we are taking the convex hull of the union of $\{x_3\}$ and S^1.

SPECIAL TYPES OF POLYTOPES AND EULER'S FORMULA

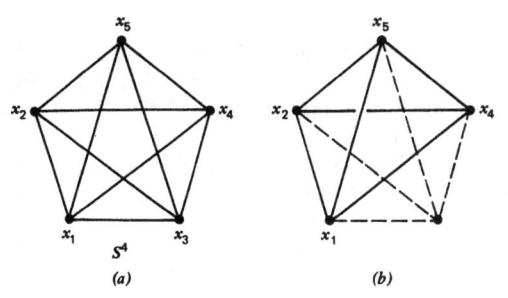

Figure 21.2.

21.2. Corollary. The number of k-faces of an n-dimensional simplex S^n in E^n is given by $f_k(S^n) = \binom{n+1}{k+1}$ where $\binom{a}{b}$ is the usual binomial coefficient $\binom{a}{b} \equiv \dfrac{a!}{b!(a-b)!}$.

In the following chart the number of proper k-faces for the first four simplices is indicated and Euler's formula for each is verified:

	$f_0(S^n)$	$f_1(S^n)$	$f_2(S^n)$	$f_3(S^n)$	$\sum_{k=0}^{n-1}(-1)^k f_k(S^n)$
S^1	2				2
S^2	3	3			0
S^3	4	6	4		2
S^4	5	10	10	5	0

In our construction of S^k we took the convex hull of S^{k-1} and a point $x \notin \operatorname{aff} S^{k-1}$. This concept may be generalized as follows.

21.3. Definition. A k-**pyramid** P^k is the convex hull of a $(k-1)$-polytope Q, called the **basis** of P^k, and a point $x \notin \operatorname{aff} Q$, called the **apex** of P^k.

21.4. Theorem. The numbers of the k-faces of an n-dimensional pyramid P^n with the $(n-1)$-polytope Q as a basis and the point x as an apex are given by

$$f_0(P^n) = f_0(Q) + 1$$

$$f_k(P^n) = f_k(Q) + f_{k-1}(Q) \quad \text{for } 1 \leq k \leq n-2.$$

$$f_{n-1}(P^n) = 1 + f_{n-2}(Q)$$

PROOF. Let F be a k-face of P^n corresponding to the hyperplane H, so that $F = H \cap P^n$. Since $\operatorname{vert} F \subset \operatorname{vert} P^n = \operatorname{vert} Q \cup \{x\}$, we have two possibilities:

(a) $x \notin \text{vert } F$. Then Theorem 20.11 implies that F is a k-face of Q.
(b) $x \in \text{vert } F$. Then the vertices of F other than x are precisely the set of vertices of the $(k-1)$-face $H \cap Q$ of Q. Thus F is the k-pyramid with basis $F \cap Q$ and apex x.

On the other hand, Theorem 20.12 implies that every face of Q (including Q itself) is a proper face of P^n. Also, the convex hull of the union of a proper $(k-1)$-face of Q and x is a proper k-face of P^n. Indeed, if F is a proper $(k-1)$-face of Q, then there exists an $(n-2)$-dimensional flat H_0 in aff Q which supports Q and such that $H_0 \cap Q = F$. But then $H \equiv \text{aff}(H_0 \cup \{x\})$ supports P^n such that $H \cap P = \text{conv}(F \cup \{x\})$. (See Figure 21.3.) ∎

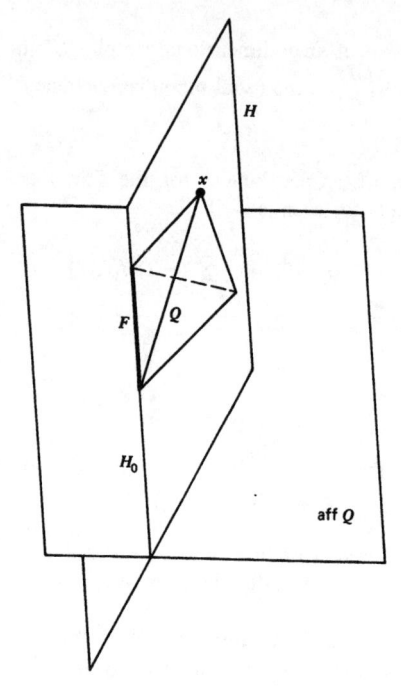

Figure 21.3.

Crosspolytopes and Bipyramids

21.5. Definition. Suppose $x_1, \ldots, x_k$ are linearly independent vectors in E^n ($1 \leq k \leq n$). Then $X^k \equiv \text{conv}\{\pm x_1, \ldots, \pm x_k\}$ is called a *k*-**crosspolytope**.

The set X^1 is a line segment symmetric about the origin. By joining points x_2 and $-x_2$ (not in the line of X^1) we obtain a parallelogram X^2. (See Figure

SPECIAL TYPES OF POLYTOPES AND EULER'S FORMULA

Figure 21.4.

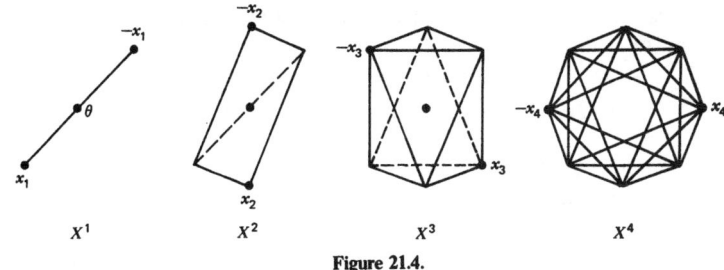

21.4.) By joining points x_3 and $-x_3$ (not in the plane of X^2) we obtain an octahedron X^3. Our drawing of X^4 in Figure 21.4 is suggestive of the four-dimensional crosspolytope. We have joined points x_4 and $-x_4$ (not in the three-space of X^3!) to obtain a figure with 8 vertices and 16 facets. (See Exercise 21.1.)

Just as the simplex is a special case of the pyramid, the crosspolytope is a special case of the bipyramid.

21.6. Definition. Let Q be a $(k-1)$-polytope and let I be a closed line segment such that relint $Q \cap$ relint I is a single point. Then the k-polytope $P = \operatorname{conv}(Q \cup I)$ is called a **k-bipyramid** with **basis** Q.

In this definition, if $I = \overline{xy}$, then by reasoning analogous to that for the pyramids we see that each face of P is one of the following:
  (a) A proper face of Q.
  (b) A pyramid with a proper face of Q as basis, and either x or y as apex.
  (c) One of the vertices $\{x\}$ or $\{y\}$.

From this observation we obtain the following:

21.7. Theorem. Let P be an n-dimensional bipyramid in E^n having basis Q. Then the numbers of the k-faces are given by

$$f_0(P) = f_0(Q) + 2$$
$$f_k(P) = f_k(Q) + 2f_{k-1}(Q) \quad \text{for } 1 \leq k \leq n-2.$$
$$f_{n-1}(P) = 2f_{n-2}(Q)$$

From our construction of the crosspolytope, we see that each of its proper faces is a simplex. This observation plus an induction argument yield the following corollary, the details being left to the reader (Exercise 21.3).

21.8. Corollary. The numbers of the k-faces of the n-crosspolytope X^n in E^n are given by

$$f_k(X^n) = 2^{k+1}\binom{n}{k+1}, \qquad 0 \leq k \leq n-1.$$

In the following chart the number of proper k-faces for each crosspolytope in Figure 21.4 is indicated and Euler's formula for each is verified.

	$f_0(X^n)$	$f_1(X^n)$	$f_2(X^n)$	$f_3(X^n)$	$\sum_{k=0}^{n-1}(-1)^k f_k(X^n)$
X^1	2				2
X^2	4	4			0
X^3	6	12	8		2
X^4	8	24	32	16	0

Parallelotopes and Prisms

21.9. Definition. Suppose $x_1, \ldots, x_k$ are linearly independent vectors in E^n ($1 \leq k \leq n$). Let $I_j \equiv \overline{\theta x_j}$ be the line segment from the origin θ to x_j ($j = 1, \ldots, k$). Then the vector sum

$$P^k \equiv I_1 + I_2 + \cdots + I_k$$

is called a **k-parallelotope**. If the vectors $x_1, \ldots, x_k$ are mutually orthogonal unit vectors, then P^k is called a **hypercube**.

The parallelotope P^1 is a line segment. If P^1 is translated by a vector x_2 (not in the line of P^1), its initial and final positions determine a parallelogram P^2. (See Figure 21.5.) Translating P^2 by a vector x_3 (not in the plane of P^2) determines a parallelepiped P^3. A similar translation of P^3 by a vector x_4 (not in the three-space of P^3) yields P^4.

From the definition it follows that $P^n = P^{n-1} + I_n$. Each addition of this sort doubles the number of vertices, so that P^n has 2^n vertices. Clearly, P^n can also be written as the convex hull of P^{n-1} and its translate $P^{n-1} + x_n$. Thus we see that a proper k-face of P^n is either a k-face of P^{n-1} or of $x_n + P^{n-1}$, or it is the vector sum of I_n with some $(k-1)$-face of P^{n-1}.

Conversely, each face of P^{n-1} (including itself) is a face of P^n, and the same is true for faces of $x_n + P^{n-1}$. Also, the vector sum of I_n with any face of P^{n-1} is a face of P^n. Thus we conclude that the number of k-faces of P^n is given by

$$f_k(P^n) = 2f_k(P^{n-1}) + f_{k-1}(P^{n-1}), \qquad 1 \leq k \leq n-1.$$

Using induction, we obtain the following theorem.

SPECIAL TYPES OF POLYTOPES AND EULER'S FORMULA 129

Figure 21.5.

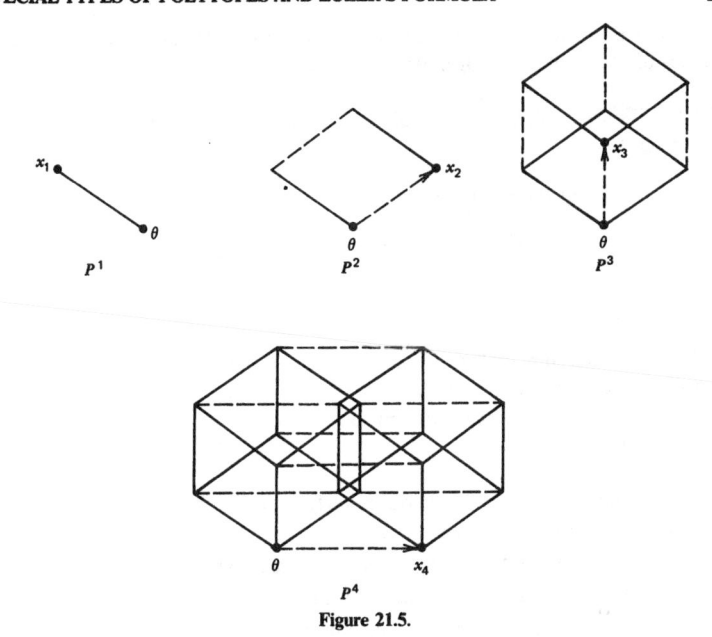

21.10. Theorem. The number of k-faces of the n-parallelotope P^n in E^n is given by

$$f_k(P^n) = 2^{n-k}\binom{n}{k}, \quad 0 \leq k \leq n.$$

If we compute the number of proper k-faces in the first four parallelotopes, we obtain the values given in the following chart:

	$f_0(P^n)$	$f_1(P^n)$	$f_2(P^n)$	$f_3(P^n)$	$\sum_{k=0}^{n-1}(-1)^k f_k(P^n)$
P^1	2				2
P^2	4	4			0
P^3	8	12	6		2
P^4	16	32	24	8	0

Following the pattern of our earlier generalizations, we may generalize the parallelotope to get a prism.

21.11. Definition. Let Q be a $(k-1)$-polytope in E^n and let $I = \overline{\theta x}$ be a line segment not parallel to aff Q. Then the vector sum $P = Q + I$ is called a **k-prism** with **basis** Q.

With a proof analogous to that used for parallelotopes, we may count the k-faces of P to obtain the following theorem.

21.12. Theorem. The number of proper k-faces of the n-prism P with basis Q is given by

$$f_0(P) = 2f_0(Q)$$

$$f_k(P) = 2f_k(Q) + f_{k-1}(Q) \qquad 1 \leq k \leq n-1.$$

Euler's Formula

Section 21 concludes with a proof of Euler's formula in three-space. For the n-dimensional case the reader is referred to the books by Coxeter (1963) and Grünbaum (1967). In developing our proof it will be helpful to use the concept of a network.

21.13. Definition. A **network** on a surface in E^3 is a finite set of points (called **nodes**) and line segments such that
  (a) Each node is the endpoint of a line segment.
  (b) Each line segment has two nodes.
  (c) Two line segments can intersect only at their endpoints. A network divides the surface into **regions**, where each region is bounded by line segments.

There are many examples and applications of networks, but we will only consider **polyhedral networks** where the surface is the boundary of a polytope P in E^3, the nodes are vertices, the line segments are edges, and the regions are facets (or the union of facets, if not all the edges are included). If we let v, e, r denote the number of vertices, edges, and regions respectively, then the expression $v - e + r$ is called the Euler characteristic of such a network.

Beginning at any vertex x of a polytope P, we can build a network to cover P in the following manner: Adjoin to x all the edges which contain x. Call this network N_1. To N_1 we now adjoin all the edges that intersect it and call it N_2. To N_2 we adjoin all the edges that intersect it and call it N_3. After a finite number of steps, we have a network that includes all the vertices and edges of P.

This building process involves additions of two types:

Type 1 The edge adjoined adds a vertex not already in the network (e.g., edge e_5 in Figure 21.6).

Type 2 The edge adjoined connects two vertices already in the network (e.g., edge e_7 in Figure 21.6).

We observe that both types of additions leave the Euler characteristic of a network unchanged.

SPECIAL TYPES OF POLYTOPES AND EULER'S FORMULA

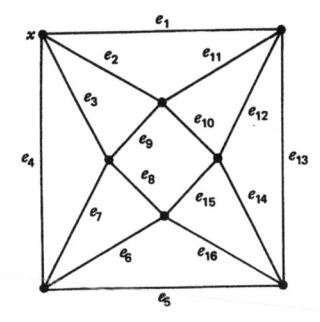

Figure 21.6. $N_1 = \{e_1, \ldots, e_4\}$; $N_2 = \{e_1, \ldots, e_{13}\}$; $N_3 = \{e_1, \ldots, e_{16}\}$.

21.14. Lemma. Let N be a polyhedral network. Additions to N of Type 1 and Type 2 do not change the Euler characteristic $v - e + r$ of N.

PROOF. Let N' (N'', respectively) be a network obtained from N by adjoining one edge of Type 1 (Type 2, respectively). If v, e, r denote the number of vertices, edges, and regions of N, then for the network N' these numbers are $v + 1$, $e + 1$, and r, respectively. For the network N'' we have v, $e + 1$, and $r + 1$, respectively. In both cases the Euler characteristic $v - e + r$ is unchanged. ∎

Using this lemma, we are now able to prove Euler's Theorem for polytopes in $\mathbb{E}^3$.

21.15. Theorem (Euler). Let P be a polytope in $\mathbb{E}^3$ and let v, e, f denote the number of vertices, edges and facets (respectively) of P. Then $v - e + f = 2$.

PROOF. Let x be a vertex of P. The polyhedral network N consisting of x alone has one vertex, no edges, and one region (the whole surface). Thus its Euler characteristic is two. By a finite number of additions of Type 1 and Type 2, we can build N into a network N' that includes all the vertices and edges of P. By Lemma 21.14, these additions do not change the Euler characteristic, so N' also has its Euler characteristic equal to two. Since the regions of N' correspond to the facets of P, we have $v - e + f = 2$. ∎

The preceding argument is disarmingly (perhaps deceptively) simple, and there is one subtlety that warrants further comment. We stated that adjoining one edge of Type 2 to a network increases the number of regions by one. This is true for a polyhedral network because the polytope underlying the network is convex and thus has no "holes" in it. This implies that every closed path divides the surface into two regions. In the nonconvex picture frame in Figure 21.7, we see that the closed path $\overline{x_1 x_2} \cup \overline{x_2 x_3} \cup \overline{x_3 x_1}$ loops through the hole and thus only determines one region on the surface. That is, adjoining edge $\overline{x_2 x_3}$ to

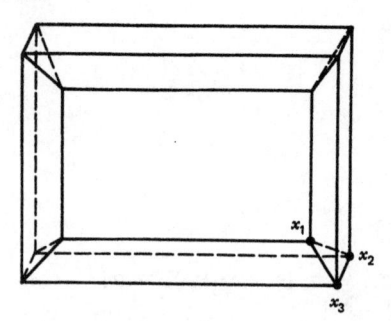

Figure 21.7.

the network $N \equiv \{\overline{x_1 x_2}, \overline{x_1 x_3}\}$ does not increase the number of regions. Since a convex polytope is simple (i.e., it has no "holes"), this situation cannot occur.

EXERCISES

21.1. In Figure 21.4, label the vertices of the crosspolytope X^4 clockwise starting at the top by a, b, c, d, e, f, g, h. By listing the vertices of each set, describe the 16 facets.

21.2. In Figure 21.5, label the vertices of P^4 in the following manner: Let a, b, c, d be the vertices of P^2 labeled clockwise from the top. Let e, f, g, h, be their translated images (respectively) in P^3. Then let $a', b', \ldots, h'$ be the translated images of the vertices of P^3 in P^4. By listing the vertices in each set, describe the 8 facets of P^4.

21.3. Complete the proof of Corollary 21.8.

21.4. Complete the proof of Theorem 21.10.

21.5. Compute $f_k(S^5)$ for $k = 0, \ldots, 4$ and verify Euler's formula.

21.6. Define $f_{-1}(P) \equiv 1$ and $f_n(P) \equiv 1$, corresponding to the two improper faces of an n-polytope P.

(a) Show that Euler's formula becomes

$$\sum_{k=-1}^{n} (-1)^k f_k(P) = 0.$$

(b) Prove that the sum of the coefficients in the polynomial expansion of $(1 - x)^n$ is equal to zero.

(c) Use parts (a) and (b) to prove Euler's formula for the n-simplex S^n.

21.7. A polyhedron (3-polytope) is called **regular** if all its facets are congruent regular polygons and all the angles at the vertices are equal. Use Euler's Theorem 21.15 to show that there are only five regular polyhedra.

APPROXIMATIONS BY POLYTOPES

*21.8. Let P be a k-polytope ($k \geq 3$) and let x be a vertex of P. If the midpoints of all the edges of P that contain x lie in a hyperplane, then these midpoints are the vertices of a $(k-1)$-dimensional polytope called the **vertex figure** of P at x. By using the usual definition of a regular polygon in E^2, we can define a regular polytope inductively as follows: The k-polytope P ($k \geq 3$) is **regular** if each of its facets is regular and there is a regular vertex figure at each vertex.
  (a) If P is a regular polytope, prove that its facets are all congruent and its vertex figures are all congruent.
  (b) Prove that there are exactly six regular 4-polytopes and three regular k-polytopes for each $k \geq 5$.

SECTION 22. APPROXIMATIONS BY POLYTOPES

One of the important applications of polytopes is their use in approximating convex sets. As we have already observed, this leads to natural definitions of volume and surface area. In Section 11, we defined the perimeter and area of planar sets by means of inscribed polygons (Definitions 11.9 and 11.10). In this section we shall define n-dimensional volume and surface area by means of circumscribing polytopes and observe that these approaches are equivalent. We begin with a definition.

22.1. Definition. A convex subset S of E^n is called a **convex body** if int $S \neq \emptyset$.

If P is an n-polytope in E^n ($n \geq 2$), then we let $V(P)$ denote the volume of P and $A(P)$ denote the surface area of bd P in the sense of the usual Euclidean measure. [When $n = 2$, $V(P)$ refers to the two-dimensional area and $A(P)$ refers to the perimeter of P.] Since the surface area of a facet F of P is the same as the volume of F considered as an $(n-1)$-polytope in aff F, these definitions are consistent. Using them we now define the volume and surface area of an arbitrary compact convex body.

22.2. Definition. Let S be a compact convex body in E^n. The volume and surface area of S are defined, respectively, by

$$V(S) \equiv \inf_{P \supset S} V(P)$$

$$A(S) \equiv \inf_{P \supset S} A(P),$$

where P is a polytope.

If the convex set S in Definition 22.2 is itself a polytope, then we might have two different meanings for $V(S)$: the usual Euclidean measure and the infimum given by the definition. Fortunately, this does not cause a problem since the values obtained are the same. Indeed, suppose Q is a polytope and let V_E denote Euclidean volume. For any polytope P containing Q, we have $V_E(P) \geq V_E(Q)$. But Q is also a polytope containing Q, and so it follows that

$$V(Q) \equiv \inf_{P \supset Q} V_E(P) = V_E(Q).$$

Similar comments apply to the definition of $A(S)$.

To verify that Definition 22.2 is consistent with our earlier definitions for planar sets, we need to be able to approximate a compact convex body S by a polytope. In the following lemma we show that there are polytopes inside S that are arbitrarily "close" to S, and in the subsequent theorem we show that S can be sandwiched between two polytopes which are arbitrarily "close" to each other. The measure of closeness used here is based on parallel bodies and the distance function described in Section 14. Our development follows the approach used by Valentine (1964).

22.3. Lemma. If S is a compact convex body in E^n, then for each $\varepsilon > 0$ there exists a polytope P such that

$$P \subset S \subset P_\varepsilon,$$

where P_ε is the parallel body to P. (See Definition 14.1.)

PROOF. Since S is compact, for each $\varepsilon > 0$ there exist finitely many points $x_1, \ldots, x_k$ in S such that

$$\text{bd } S \subset \bigcup_{i=1}^{k} B(x_i, \varepsilon) \tag{1}$$

(see Definition 1.7). Let $P = \text{conv}\{x_1, \ldots, x_k\}$. Since each x_i is in the convex set S, we have $P \subset S$. From (1) it follows that $S \subset P_\varepsilon$. ∎

22.4. Theorem. Let S be a compact convex body in E^n and suppose $\theta \in \text{int } S$. Then for each $\delta > 1$ there exists a polytope P such that

$$P \subset S \subset \delta P, \tag{2}$$

where $\delta P \equiv \{\delta x : x \in P\}$.

PROOF. Since $\theta \in \text{int } S$, there exists $r > 0$ such that the closed ball

$$K \equiv \{x : \|x\| \leq r\}$$

is contained in int S. Choose $\varepsilon > 0$ so that

$$0 < \varepsilon < r(\delta - 1). \tag{3}$$

APPROXIMATIONS BY POLYTOPES

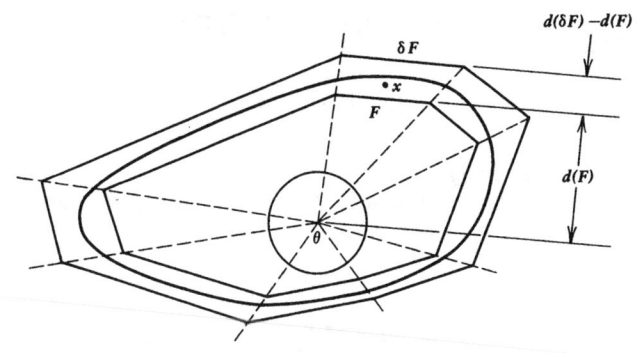

Figure 22.1.

By Lemma 22.3 there exists a polytope P such that

$$K \subset P \subset S \subset P_\varepsilon. \tag{4}$$

Let F and δF be two parallel facets of P and δP, respectively. (See Figure 22.1.) If $d(F)$ denotes the distance from the origin θ to the hyperplane containing F, then we have

$$d(\delta F) = \delta d(F),$$

so that the distance between F and δF is given by

$$d(\delta F) - d(F) = (\delta - 1)d(F). \tag{5}$$

Now let $x \in P_\varepsilon \sim P$ and let F be a facet of P which is closest to x. Then by (3) and (4) we have

$$\inf_{y \in F} \|x - y\| \leq \varepsilon < r(\delta - 1) < (\delta - 1)d(F),$$

so that $x \in \delta P$ by (5). Thus $P_\varepsilon \subset \delta P$, and this, together with (4), implies (2). ∎

22.5. Theorem. Let S be a compact convex body in E^n and define

$$U(S) \equiv \sup_{P \subset S} V(P)$$

$$B(S) \equiv \sup_{P \subset S} A(P),$$

where P is a polytope. Then $U(S) = V(S)$ and $B(S) = A(S)$.

PROOF. If P_1 and P_2 are any polytopes with $P_1 \subset S \subset P_2$, then $V(P_1) \leq V(P_2)$. It follows that

$$U(S) \equiv \sup_{P \subset S} V(P) \leq \inf_{P \supset S} V(P) \equiv V(S).$$

Similarly, we find that $B(S) \leq A(S)$.

We now assume without loss of generality that $\theta \in \operatorname{int} S$ and choose $\delta > 1$. By Theorem 22.4 there exists a polytope Q such that

$$Q \subset S \subset \delta Q.$$

It follows readily from the definition of Euclidean volume and surface area that for the polytope Q we have

$$V(\delta Q) = \delta^n V(Q)$$

and

$$A(\delta Q) = \delta^{n-1} A(Q).$$

Since $Q \subset S \subset \delta Q$, we have

$$V(Q) \leq \sup_{P \subset S} V(P) \equiv U(S) \leq V(S) \equiv \inf_{P \supset S} V(P) \leq V(\delta Q)$$

and

$$A(Q) \leq \sup_{P \subset S} A(P) \equiv B(S) \leq A(S) \equiv \inf_{P \supset S} A(P) \leq A(\delta Q).$$

Combining these properties, we obtain

$$|V(S) - U(S)| \leq (\delta^n - 1)V(Q)$$

and

$$|A(S) - B(S)| \leq (\delta^{n-1} - 1)A(Q).$$

Since this holds for any $\delta > 1$, we must have $U(S) = V(S)$ and $B(S) = A(S)$. ∎

We are now in a position to prove that the volume and surface area functions are continuous. (See Definition 16.3.) This will complete our answer to the question of the existence of extremal sets (Section 16) and the isoperimetric problem (Section 13).

22.6. Theorem. Let $\mathcal{F}$ be a family of compact convex bodies in E^n. Then the volume and surface area functions V and A are continuous on $\mathcal{F}$.

EXERCISES

PROOF. Suppose $S \in \mathcal{F}$ and assume without loss of generality that $\theta \in \text{int } S$. We shall prove that V is continuous, the proof for A being similar.

Given any $\varepsilon > 0$ there exist $\delta > 1$ and a polytope P with $P \subset S \subset \delta P$ such that

$$V(\delta P) - V(P) < \varepsilon.$$

But for any convex body T satisfying $P \subset T \subset \delta P$ we have $V(P) \leq V(T) \leq V(\delta P)$. Thus

$$|V(T) - V(S)| < \varepsilon.$$

This implies that $\lim_{i \to \infty} V(T_i) = V(S)$ for any sequence $\{T_i\}$ that converges to S. ∎

We conclude this section with an approximation theorem that will be useful to us in studying the analytic properties of convex functions (Section 30). It is also interesting in its own right.

22.7. Theorem. Let S be a convex set and suppose T is a nonempty compact subset of relint S. Then there exists a polytope P such that $T \subset P \subset S$.

PROOF. We suppose without loss of generality that S is n-dimensional so that relint $S = \text{int } S$. Let $x \in T$. Since $T \subset \text{int } S$, there exists an open ball B_x around x that is contained in S. Let P_x be a simplex such that

$$x \in \text{int } P_x \subset P_x \subset B_x \subset S.$$

Then the collection $\{\text{int } P_x : x \in T\}$ forms an open covering for T and by the Heine-Borel Theorem 1.21, there exist finitely many points $x_1, \ldots, x_m$ in T such that $T \subset \bigcup_{i=1}^{m} \text{int } P_{x_i}$. Now let $P \equiv \text{conv } \bigcup_{i=1}^{m} P_{x_i}$. Since each P_{x_i} is a simplex, P has finitely many extreme points and thus is a polytope. Since each P_{x_i} is contained in the convex set S, so is the convex hull of their union. Thus $T \subset P \subset S$, as required. ∎

EXERCISES

22.1. Let S be a circle centered at θ in E^2 and let P be a regular k-gon inscribed in S. Find the smallest $\delta > 1$ so that $P \subset S \subset \delta P$.

22.2. Let S be a circle centered at θ in E^2 and let $\delta > 1$. Find the smallest positive integer k so that the regular k-gon P inscribed in S satisfies $P \subset S \subset \delta P$.

22.3. Let S be a circle centered at θ in E^2. Find the smallest positive integers k_i ($i = 1, \ldots, 4$) so that the regular k_i-gon P_i inscribed in S satisfies $P_i \subset S \subset \delta_i P_i$, where $\delta_1 = 2$, $\delta_2 = 1.5$, $\delta_3 = 1.2$, and $\delta_4 = 1.05$.

Note. Exercises 22.4–22.7 outline one way to verify the assertion in the proof of Theorem 22.5 that for an n-polytope P in E^n we have $V(\delta P) = \delta^n V(P)$. The following definition is useful.

Definition. Let P be an n-polytope in E^n ($n \geq 2$). Then P is **decomposed** into n-polytopes $P_1, \ldots, P_k$ iff $P = \bigcup_{i=1}^{k} P_i$ and int $P_i \cap$ int $P_j = \emptyset$ for $i \neq j$.

22.4. Prove that any n-polytope P in E^n ($n \geq 2$) can be "triangulated." That is, prove that P can be decomposed into a finite number of n-simplices.

*22.5. Let P be an n-simplex in E^n ($n \geq 2$). Prove that $V(P) = A(Q) \cdot h/n$, where Q is any one of the facets of P, and h is the corresponding altitude from Q to the opposite vertex.

22.6. Let P be an n-simplex in E^n ($n \geq 2$) and let $\delta > 1$. Use Exercise 22.5 to show that $V(\delta P) = \delta^n V(P)$.

22.7. If the n-polytope P is decomposed into n-polytopes $P_1, \ldots, P_k$, then the n-polytope δP (where $\delta > 1$) is decomposed into the n-polytopes $\delta P_1, \ldots, \delta P_k$. Furthermore, $V(P) = \Sigma_{i=1}^{k} V(P_i)$, so that $V(\delta P) = \Sigma_{i=1}^{k} V(\delta P_i)$. Use these facts and the preceding exercise to verify $V(\delta P) = \delta^n V(P)$ for any n-polytope P.

*22.8. Two polygons P and Q in E^2 are called **equidecomposable** if there is a polygonal decomposition $P_1, \ldots, P_k$ of P and a polygonal decomposition $Q_1, \ldots, Q_k$ of Q such that P_i is congruent to Q_i for $i = 1, \ldots, k$. Prove that any two polygons of equal area are equidecomposable. (Note: It is not necessary that these polygons be convex, only that their boundaries consist of a finite number of line segments.)

22.9. Is the converse of Exercise 22.8 true? That is, if two polygons are equidecomposable, must they have the same area? Comment on the sets in Figure 22.2, where P_i is congruent to Q_i for $i = 1, \ldots, 7$.

22.10. Two sets A and B are said to be **equivalent by finite decomposition** if A is the union of disjoint sets $A_1, \ldots, A_m$ and B is the union of disjoint sets $B_1, \ldots, B_m$ such that A_i is congruent to B_i for all i ($i = 1, \ldots, m$).

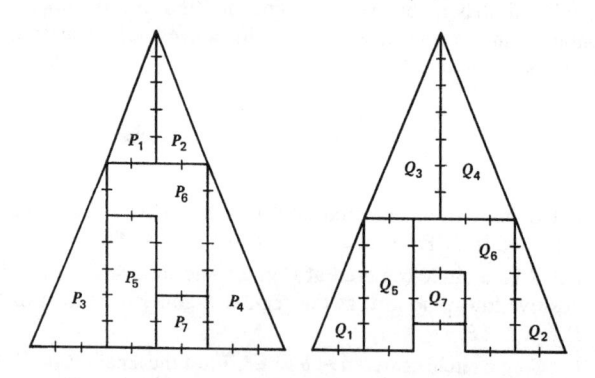

Figure 22.2.

EXERCISES

The sets A_i are called the **factors** of the decomposition. If each factor is a convex set, we say that A and B are **convex equidecomposable**.

†(a) Suppose A and B are compact, convex sets in E^2 which are equivalent by finite decomposition. Are A and B also convex equidecomposable?

†(b) Is it possible for a circular region and a square in E^2 to be equivalent by finite decomposition?

9

DUALITY

In many different branches of mathematics one encounters the term *duality*. In general, this refers to two similar ideas or theories that are related to each other, usually in an inverse way. In our study of convex sets and their applications we shall encounter three examples of duality—two will be given in this chapter and one will be found later in Section 26.

In Section 23 we introduce the idea of the polar of a set and look at the application of this to polytopes. In Section 24 we define the dual cone of a set and use it to derive alternative proofs for some earlier theorems. In both cases the inverse relationship changes unions to intersections and reverses set inclusion. For example, if we use an asterisk to denote the polar of a set, then we will show that

$$A \subset B \text{ implies } B^* \subset A^*$$

and

$$(A \cup B)^* = A^* \cap B^*.$$

When we get to Section 26 and encounter duality in the context of linear programming, the inverse relationship will change a maximizing problem into a minimizing problem.

SECTION 23. POLARITY AND POLYTOPES

23.1. Definition. Let K be any nonempty subset of E^n. Then the **polar set** K^* of K is defined by

$$K^* \equiv \{y \in \mathsf{E}^n : \langle x, y \rangle \leq 1 \text{ for all } x \in K\}.$$

Before deriving some of the basic properties of a polar set, let us look at a couple of examples.

23.2. Examples

(a) If K consists of a single point, say x, and $x \neq \theta$, then K^* is the closed half-space which is bounded by the hyperplane

$$\{y: \langle x, y \rangle = 1\}$$

and which contains the origin in its interior.

(b) If K consists just of the origin, then $K^* = \mathsf{E}^n$, since for any $y \in \mathsf{E}^n$ we have

$$\langle \theta, y \rangle = 0 \leq 1.$$

(c) If $S(r, \theta)$ is a closed ball of radius $r > 0$ centered at θ, then

$$[S(r, \theta)]^* = S\left(\frac{1}{r}, \theta\right).$$

To see this, we recall from linear algebra (see Theorem 1.4) that

$$\langle x, y \rangle = \|x\| \|y\|$$

whenever x lies on the ray $R(\theta, y)$ from the origin through y. Thus if $y \in [S(r, \theta)]^*$, then by letting x' be the point where the ray $R(\theta, y)$ intersects bd $S(r, \theta)$ we have

$$1 \geq \langle x', y \rangle = \|x'\| \|y\| = r \|y\|,$$

whence $\|y\| \leq 1/r$ and so $y \in S(1/r, \theta)$. (See Figure 23.1.)

Conversely, if $z \in S(1/r, \theta)$, then for any $x \in S(r, \theta)$ we have

$$\langle x, z \rangle = \|x\| \|z\| \cos \gamma \leq r\left(\frac{1}{r}\right)(1) = 1,$$

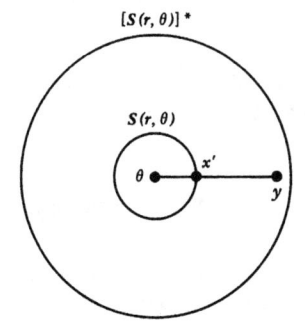

Figure 23.1.

where γ is the angle between $\overline{\theta x}$ and $\overline{\theta z}$. Thus $z \in [S(r, \theta)]^*$ and we conclude $[S(r, \theta)]^* = S(1/r, \theta)$.

23.3. Theorem. Let K, K_1, K_2, and K_α ($\alpha \in \mathcal{Q}$) be nonempty sets. Then the following are true:
(a) $[\bigcup_{\alpha \in \mathcal{Q}} K_\alpha]^* = \bigcap_{\alpha \in \mathcal{Q}} (K_\alpha)^*$.
(b) K^* is a closed convex set containing θ.
(c) If $K_1 \subset K_2$, then $K_2^* \subset K_1^*$.
(d) If $\lambda > 0$, then $(\lambda K)^* = (1/\lambda) K^*$.

PROOF.
(a) From the definition of a polar set we have

$$\left[\bigcup_{\alpha \in \mathcal{Q}} K_\alpha \right]^* = \left\{ y \in \mathsf{E}^n : \langle x, y \rangle \leq 1 \text{ for all } x \in \bigcup_{\alpha \in \mathcal{Q}} K_\alpha \right\}$$

$$= \bigcap_{\alpha \in \mathcal{Q}} \{ y \in \mathsf{E}^n : \langle x, y \rangle \leq 1 \text{ for all } x \in K_\alpha \}$$

$$= \bigcap_{\alpha \in \mathcal{Q}} (K_\alpha)^*.$$

(b) We saw in Example 23.2 that the polar set of a single point is either E^n or a closed half-space. From part (a) we have

$$K^* = \left[\bigcup_{x \in K} \{x\} \right]^* = \bigcap_{x \in K} \{x\}^*,$$

so that K^* is the intersection of a collection of closed convex sets. Thus K^* is closed and convex. Since each $\{x\}^*$ contains θ, so does K^*.
(c) If $K_1 \subset K_2$, then $K_2 = K_1 \cup (K_2 \sim K_1)$. But then $K_2^* = [K_1^* \cap (K_2 \sim K_1)^*] \subset K_1^*$ by part (a).
(d) If $x \in (\lambda K)^*$, then $\langle \lambda k, x \rangle \leq 1$ for all $\lambda k \in \lambda K$. That is, $\langle k, \lambda x \rangle \leq 1$ for all $k \in K$, which implies $\lambda x \in K^*$ and $x \in (1/\lambda) K^*$. The converse is similar. ∎

Although these properties of polar sets apply to any nonempty subsets of E^n, by placing restrictions on the sets we may obtain the stronger results in the following two theorems.

23.4. Theorem. Let K be a compact convex set that contains the origin as an interior point. Then K^* is compact and $\theta \in \text{int } K^*$.

PROOF. Since $\theta \in \text{int } K$, there exists $r > 0$ such that the closed ball $S(r, \theta)$ centered at θ with radius r is contained in K. In Example 23.2(c) we saw that

POLARITY AND POLYTOPES

$[S(r, \theta)]^* = S(1/r, \theta)$. Thus from Theorem 23.3(c) we have $K^* \subset S(1/r, \theta)$, so K^* is bounded. But part (b) of Theorem 23.3 says that K^* is closed, so K^* is also compact. Similarly, since K is bounded, there exists $R > 0$ such that $K \subset S(R, \theta)$. But then $S(1/R, \theta) \subset K^*$, so $\theta \in \operatorname{int} K^*$. ∎

23.5. Theorem. Let K be a closed convex set containing the origin. Then $K^{**} = K$, where $K^{**} \equiv (K^*)^*$.

PROOF. If $x \in K$, then $\langle x, y \rangle \leq 1$ for all $y \in K^*$. Thus $x \in K^{**}$ and $K \subset K^{**}$. Conversely, if $x_0 \notin K$, then by Theorem 4.12 there exists a hyperplane

$$H \equiv \{z \in \mathsf{E}^n : \langle z, u \rangle = 1\}$$

strictly separating x_0 and K. Since $\theta \in K$, we have $\langle x, u \rangle < 1$ for all $x \in K$ and $\langle x_0, u \rangle > 1$. The first inequality implies that $u \in K^*$, and so the second inequality implies $x_0 \notin K^{**}$. Hence $K^{**} \subset K$, and we conclude that $K^{**} = K$. ∎

Having derived some of the important properties of polar sets, we now look at duality in polytopes and apply the theory of polar sets to prove that each polytope has a dual. To be more precise, we need the following definition.

23.6. Definition. Let P be a polytope. Then a polytope P^* is **dual** to P if there exists a one-to-one correspondence between the set of faces of P and the set of faces of P^* that reverses the relation of inclusion.

In this definition we have used the notation P^* to represent a polytope dual to P. This is the same notation as we have been using for the polar set of P. This usage will be justified by Theorem 23.11 when we prove that if P is a polytope with $\theta \in \operatorname{int} P$, then the polar set of P is itself a polytope and is dual to P.

23.7. Example. The m-sided prism and the m-sided bipyramid are dual in E^3. When $m = 4$, we have the cube and the octahedron (illustrated in Figure 23.2). The face F_1 corresponds to the vertex V_1'. The edges E_i ($i = 1, \ldots, 4$) correspond to the edges E_i' ($i = 1, \ldots, 4$). The vertex V_1 corresponds to the face F_1'. Note that $V_1 \subset E_3 \subset F_1$, while $F_1' \supset E_3' \supset V_1'$. Thus the correspondence reverses set inclusion.

23.8. Definition. Let K be a compact convex set with $\theta \in \operatorname{int} K$. For any face F of K, proper or improper, we define the subset $\hat{F}$ of K^* by

$$\hat{F} \equiv \{y \in K^* : \langle y, x \rangle = 1 \text{ for all } x \in F\}.$$

23.9. Lemma. $\hat{F}$ is a face of K^*.

144 DUALITY

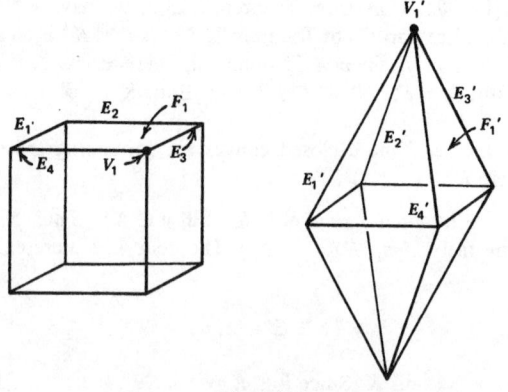

Figure 23.2.

PROOF. Since $\hat{\phi} = K^*$ and $\hat{K} = \emptyset$, we have only to consider the case where F is a proper face of K. Choose $x_0 \in \text{relint } F$ and let $F' \equiv H \cap K^*$, where H is the supporting hyperplane to K^* given by

$$H \equiv \{y \in \mathsf{E}^n : \langle y, x_0 \rangle = 1\}.$$

Then F' is a face of K^*, and clearly $\hat{F} \subset F'$.

We claim that in fact $\hat{F} = F'$. To see this, suppose $y_0 \in K^* \sim \hat{F}$. Then there exists $x_1 \in F$ such that $\langle y_0, x_1 \rangle < 1$. Since $x_0 \in \text{relint } F$, there exists $x_2 \in F$ such that

$$x_0 = (1 - \lambda)x_1 + \lambda x_2$$

for some λ satisfying $0 < \lambda < 1$ (see Figure 23.3). Since $y_0 \in K^*$ we have $\langle y_0, x_2 \rangle \leq 1$ and it follows that

$$\langle y_0, x_0 \rangle = (1 - \lambda)\langle y_0, x_1 \rangle + \lambda \langle y_0, x_2 \rangle < 1.$$

Therefore $y_0 \notin F'$, and since $\hat{F} \subset F'$ we conclude that $\hat{F} = F'$. Thus $\hat{F}$ is a face of K^*. ∎

23.10. Lemma. *The map which sends each face F of K into the face $\hat{F}$ of K^* is a one-to-one mapping onto the set of all faces of K^* and reverses inclusion.*

PROOF. To prove that the mapping is one-to-one onto the set of all faces of K^*, it suffices to show that $\hat{\hat{F}} = F$. Now by definition,

$$\hat{\hat{F}} \equiv \{x \in K^{**} : \langle x, y \rangle = 1 \text{ for all } y \in \hat{F}\}.$$

POLARITY AND POLYTOPES

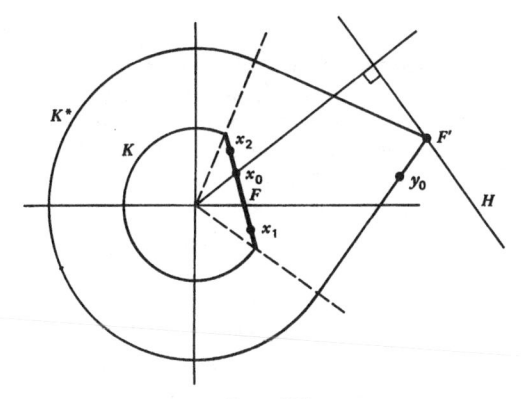

Figure 23.3.

But $K^{**} = K$, by Theorem 23.5. Thus if $x \in F$ and $y \in \hat{F}$ we have $x \in K^{**}$ and $\langle x, y \rangle = 1$. Whence $x \in \hat{\hat{F}}$ and we conclude $F \subset \hat{\hat{F}}$.

The only nontrivial case of the converse is when F is a proper face of K. In this case there exists a supporting hyperplane of K,

$$H \equiv \{x \in \mathsf{E}^n : \langle x, y_0 \rangle = 1\},$$

such that $F = H \cap K$ and $K \subset \{x : \langle x, y_0 \rangle \leq 1\}$. Since $\langle x, y_0 \rangle = 1$ for all $x \in F$, $y_0 \in \hat{F}$. But now, if $z \in K \sim F$, then $\langle z, y_0 \rangle < 1$, and so $z \notin \hat{\hat{F}}$. Thus $\hat{\hat{F}} \subset F$ and we conclude $\hat{\hat{F}} = F$.

The proof that the mapping reverses inclusion follows directly from the definition and is left as an exercise. ∎

We are now in a position to prove that each n-dimensional polytope in E^n has a dual. For the sake of simplicity in the following theorem, we assume that $\theta \in \text{int } P$. This does not handicap our result, however, since this condition can always be satisfied by making an appropriate translation.

23.11. Theorem. Let P be an n-dimensional polytope in E^n such that $\theta \in \text{int } P$. Then the polar set P^* of P is an n-dimensional polytope dual to P.

PROOF. From Lemma 23.10 we know that the correspondence between faces F of P and $\hat{F}$ of P^* is one-to-one and inclusion reversing. Thus it suffices to show that P^* is an n-dimensional polytope.

Since P is a polytope, it has a finite number of faces. Thus from our correspondence we conclude that P^* also has a finite number of faces. This implies that P^* is a polyhedral set. Since P^* is bounded (Theorem 23.4), it follows (Theorem 20.9) that P^* is a polytope. Since $\theta \in \text{int } P^*$ (Theorem 23.4), P^* is n-dimensional. ∎

EXERCISES

23.1. Describe the polar set of each subset of E^2.
  (a) The closed line segment from $(0, 0)$ to $(2, 0)$.
  (b) The closed line segment from $(0, -1)$ to $(0, 4)$.
  (c) The square with vertices at the points $(\pm 1, 0)$ and $(0, \pm 1)$.
  (d) The ray $\{(x, y): x \geq 3 \text{ and } y = 0\}$.
  (e) The ray $\{(x, y): x \geq 1 \text{ and } y = 1\}$.

23.3. Let K be a nonempty compact subset of E^n. Prove that $(\operatorname{conv} K)^* = K^*$.

23.3. Let K be a nonempty compact convex set and let Q be the profile of K. (See Definition 5.5.) Prove that $Q^* = K^*$.

23.4. (The Bipolar Theorem) Let K be a nonempty subset of E^n. Prove that $K^{**} = \operatorname{cl} \operatorname{conv}[K \cup \{\theta\}]$. In particular, when K is compact, then $K^{**} = \operatorname{conv}[K \cup \{\theta\}]$.

23.5. Let K be a nonempty subset of E^n. Prove the following:
  (a) $K^{***} = K^*$.
  (b) K is bounded iff $\theta \in \operatorname{int} K^*$.
  (c) K^* is bounded iff $\theta \in \operatorname{int} \operatorname{conv} K$.
  (d) If K is a polytope, then K^* is a polyhedral set.

23.6. Finish the proof of Lemma 23.10.

23.7. Let K_α ($\alpha \in \mathcal{C}$) be a family of nonempty compact convex sets each of which contains the origin in its interior. Prove that $(\bigcap_{\alpha \in \mathcal{C}} K_\alpha)^* = \operatorname{cl} \operatorname{conv} \bigcup_{\alpha \in \mathcal{C}} K_\alpha^*$.

23.8. Show that the n-sided pyramid in E^3 is dual to itself.

23.9. Show that the dodecahedron in E^3 is dual to the icosahedron.

23.10. Prove that an n-simplex in E^n is dual to itself.

*23.11. Let P be an n-polytope in E^n and let F be a k-face of P ($1 \leq k \leq n - 1$). Let $h_m(F)$ denote the number of m-faces of P which contain F if $m \geq k$, and the number of m-faces of P contained in F if $m \leq k$. Prove the following generalization of Euler's formula:

$$\sum_{i=k}^{n-1} (-1)^i h_i(F) = (-1)^{n-1}.$$

SECTION 24. DUAL CONES

The use of dual cones as a tool in problem solving was apparently first employed by J. Dieudonné (1941) in his proof of the Hahn–Banach theorem. Since then, many others [e.g., see Klee (1951) and Hille and Phillips (1957)] have used this technique in an occasional way. More recently, Valentine (1963) has looked carefully at dual cones and their applications, particularly to Helly-type theorems. Sandgren (1954) has also made significant contributions to their systematic study.

DUAL CONES

The duality involved in dual cones is similar in many respects to the duality of polar sets (see Section 23). For this reason, some of the easier proofs in this section will be left as exercises since they parallel closely our earlier results. We will find, however, that in one respect the dual relationship in dual cones is weaker. The dual cone of a set in E^n will be defined to be a particular subset of E^{n+1}. This increase in dimension will rule out the possibility of a result like $K^{**} = K$, which held with some restrictions for polar sets (Theorem 23.5).

24.1. Notation. We will identify the space E^{n+1} with the Euclidean product $\mathsf{R} \times \mathsf{E}^n$, where R is the first coordinate and the last n coordinates are usually grouped together. Thus we will write $(\alpha, x) \in \mathsf{R} \times \mathsf{E}^n$ and understand that α is a real number in R and x is an n-tuple in E^n. The inner product of two points in $\mathsf{R} \times \mathsf{E}^n$ will be denoted by $(\alpha, x) \cdot (\beta, y)$, which can be evaluated as

$$(\alpha, x) \cdot (\beta, y) = \alpha\beta + \langle x, y \rangle.$$

We are now ready for the basic definition.

24.2. Definition. Let M be any subset (proper or improper) of E^n. Then the **dual cone** dc M of M is given by

$$\text{dc } M \equiv \{(\alpha, x) \in \mathsf{R} \times \mathsf{E}^n : (\alpha, x) \cdot (-1, y) \leq 0 \text{ for all } y \in M\}.$$

This definition differs from the one used by Valentine (1963, 1964) and others, which is built on the concept of a "support function." The two definitions are essentially equivalent (see Exercise 29.3), but we prefer the one given here because it emphasizes the similarity with polar sets.

It may be helpful to visualize $\mathsf{R} \times \mathsf{E}^n$ with R as a vertical axis and E^n as a horizontal "hyperplane." The inner product in the definition can be written as

$$-\alpha + \langle x, y \rangle \leq 0$$

or, equivalently,

$$\alpha \geq \langle x, y \rangle.$$

Thus a point (α, x) will be in dc M if (α, x) is equal to or "above" the point (α_0, x), where $\alpha_0 = \sup_{y \in M} \langle x, y \rangle$.

24.2. Examples. If M consists of a single point, say y_0, then dc M is an "upper" closed half-space in $\mathsf{R} \times \mathsf{E}^n$ bounded below by the hyperplane

$$\{(\alpha, x): (\alpha, x) \cdot (-1, y_0) = 0\}.$$

Note that this bounding hyperplane passes through the origin $(0, \theta)$ in $\mathsf{R} \times \mathsf{E}^n$. (See Figure 24.1.)

148 DUALITY

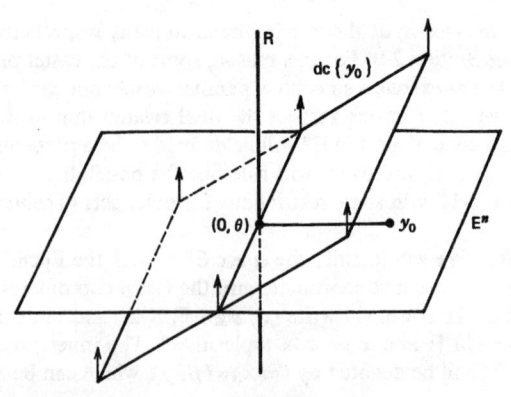

Figure 24.1.

If $M = \mathsf{E}^n$, then dc M is the ray $\{(\alpha, \theta): \alpha \geq 0\}$. Indeed, if $x \neq \theta$, then $\langle x, y \rangle$ is unbounded for $y \in M$. Thus there are no points of the dual cone above such an x. On the other hand, $\langle \theta, y \rangle = 0$ for all $y \in M$, so $(\alpha, \theta) \in$ dc M when $\alpha \geq 0$.

In the following theorem we present several of the important elementary properties of dual cones.

24.3. Theorem. Let M, N, and M_α ($\alpha \in \mathcal{C}$) be subsets of E^n. Then the following are true:

(a) $\mathrm{dc}(\bigcup_{\alpha \in \mathcal{C}} M_\alpha) = \bigcap_{\alpha \in \mathcal{C}} \mathrm{dc}\, M_\alpha$.

(b) If $M \subset N$, then dc $N \subset$ dc M.

(c) dc $M = \mathsf{R} \times \mathsf{E}^n$ if and only if $M = \varnothing$.

(d) dc $M =$ dc(cl M).

(e) dc $M =$ dc(conv M).

PROOF. The proofs of (a) and (b) are left as an exercise.

(c) If $M = \varnothing$, then the condition that $(\alpha, x) \cdot (-1, y) \leq 0$ for all $y \in M$ is vacuously satisfied for all (α, x) in $\mathsf{R} \times \mathsf{E}^n$. On the other hand, if there exists $y_0 \in M$, then dc $M \subset$ dc $\{y_0\}$ by part (b), and so dc M is contained in a half-space.

(d) Since $M \subset$ cl M, it suffices by part (b) to show that dc $M \subset$ dc(cl M). To this end, let $(\alpha, x) \in$ dc M. Then $(\alpha, x) \cdot (-1, y) \leq 0$ for all $y \in M$. Given an arbitrary point $y_0 \in$ cl M, let $\{y_n\}$ be a sequence in M that converges to y_0. Then $(\alpha, x) \cdot (-1, y_n) \leq 0$ for all n, and so

$$(\alpha, x) \cdot (-1, y) = \lim_{n \to \infty} \left[(\alpha, x) \cdot (-1, y_n) \right] \leq 0$$

by the continuity of the inner product. Thus $(\alpha, x) \in$ dc(cl M).

DUAL CONES

(e) Suppose $(\alpha, x) \in \operatorname{dc} M$. Then $\alpha \geq \langle x, y \rangle$ for all $y \in M$. Now if $z \in \operatorname{conv} M$, then by Caratheodory's Theorem 2.23 we can write $z = \sum_{i=1}^{k} \lambda_i y_i$ where $y_i \in M$, $\lambda_i \geq 0$, $\sum_{i=1}^{k} \lambda_i = 1$, and $k \leq n + 1$. It follows that

$$\langle x, z \rangle = \sum_{i=1}^{k} \lambda_i \langle x, y_i \rangle \leq \sum_{i=1}^{k} \lambda_i \alpha = \alpha.$$

Thus $(\alpha, x) \cdot (-1, z) \leq 0$ for all $z \in \operatorname{conv} M$ and so $(\alpha, x) \in \operatorname{dc}(\operatorname{conv} M)$. We conclude that $\operatorname{dc} M \subset \operatorname{dc}(\operatorname{conv} M)$. The reverse inclusion follows from part (b). ∎

We are now in a position to prove that the dual cone of a set is in fact a convex cone. (See Exercise 5.15 for the definition and basic properties of a convex cone.)

24.4. Theorem. Let M be a subset of E^n. Then $\operatorname{dc} M$ is a closed convex cone in $\mathsf{R} \times \mathsf{E}^n$ having the origin $(0, \theta)$ as its vertex and containing the ray $\{(\alpha, x): \alpha \geq 0\}$.

PROOF. If $M = \varnothing$, then $\operatorname{dc} M = \mathsf{R} \times \mathsf{E}^n$ by Theorem 24.3(c), and this clearly meets all the requirements. Thus we suppose $M \neq \varnothing$. Now the dual cone of a point is a closed half-space bounded by a hyperplane through θ, and this closed half-space is a convex cone with vertex at the origin. Theorem 24.3(a) implies that $\operatorname{dc} M$ is an intersection of such half-spaces, and it follows that $\operatorname{dc} M$ is a convex cone with vertex at the origin. Since the dual cone of E^n is the ray $\{(\alpha, \theta): \alpha \geq 0\}$, Theorem 24.3(b) implies that this ray is contained in $\operatorname{dc} M$. ∎

Up to this point we have derived properties that apply to the dual cone of any subset (proper or improper) of E^n. In the next two theorems we consider more restricted subsets, and of course obtain stronger results.

24.5. Theorem. Let M be a bounded subset of E^n. Then the ray $Q \equiv \{(\alpha, \theta): \alpha > 0\}$ is contained in the interior of $\operatorname{dc} M$. Furthermore, given any $x \in \mathsf{E}^n$, there exists some $\alpha \in \mathsf{R}$ such that $(\alpha, x) \in \operatorname{dc} M$.

PROOF. Since M is bounded, it is contained in an n-simplex S with vertices, say $y_1, \ldots, y_{n+1}$. Now $Q \subset \operatorname{int} \operatorname{dc}\{y_i\}$ $(i = 1, \ldots, n+1)$, and so

$$Q \subset \bigcap_{i=1}^{n+1} \operatorname{int} \operatorname{dc}\{y_i\} = \operatorname{int} \bigcap_{i=1}^{n+1} \operatorname{dc}\{y_i\}$$

$$= \operatorname{int} \operatorname{dc}\{y_1, \ldots, y_{n+1}\} = \operatorname{int} \operatorname{dc} S$$

$$\subset \operatorname{int} \operatorname{dc} M,$$

since $\operatorname{conv}\{y_1,\ldots,y_{n+1}\} = S$. Furthermore, if $\beta = \sup_{y \in M} \|y\|$, then given any $x \in \mathsf{E}^n$, we have $(\alpha_0, x) \in \operatorname{dc} M$ where $\alpha_0 = \beta \|x\|$. Indeed,

$$(\alpha_0, x) \cdot (-1, y) = \langle x, y \rangle - \beta \|x\|$$
$$\leq \|x\| \|y\| - \beta \|x\| \leq 0$$

for all $y \in M$. ∎

24.6. Theorem. Let M_α ($\alpha \in \mathcal{C}$) be a family of closed convex subsets of E^n. Suppose at least one of the M_α is compact. Then

$$\operatorname{dc}\left(\bigcap_{\alpha \in \mathcal{C}} M_\alpha\right) = \operatorname{cl conv}\left(\bigcup_{\alpha \in \mathcal{C}} \operatorname{dc} M_\alpha\right).$$

PROOF. For notational simplicity, we shall understand that all intersections and unions in this proof are to be taken over all α in the index set $\mathcal{C}$. Let $(\beta, z) \in \operatorname{dc}(\cap M_\alpha)$ and suppose $(\beta, z) \notin \operatorname{cl conv}(\cup \operatorname{dc} M_\alpha)$. Then by Theorem 4.12 there exists a hyperplane H_0 strictly separating (β, z) and $\operatorname{cl conv}(\cup \operatorname{dc} M_\alpha)$. Let H be the translate of H_0 which contains $(0, \theta)$. Then $(\beta, z) \notin H$ and H bounds $\operatorname{cl conv}(\cup \operatorname{dc} M_\alpha)$. By Theorem 24.5, $(1, \theta) \in \operatorname{int dc} M_\alpha$ for at least one α, and so $(1, \theta) \notin H$. Thus we can write

$$H = \{(\delta, x): (\delta, x) \cdot (-1, y_0) = 0\}$$

for some fixed $y_0 \in \mathsf{E}^n$. Since (β, z) is below H, it follows that $(\beta, z) \cdot (-1, y_0) > 0$ (i.e., $\langle z, y_0 \rangle > \beta$). But $\beta \geq \sup_{x \in \cap M_\alpha} \langle z, x \rangle$ since $(\beta, z) \in \operatorname{dc}(\cap M_\alpha)$. Thus $y_0 \notin \cap M_\alpha$. On the other hand, for each α we have

$$\langle y_0, w \rangle \leq \sup_{y \in M_\alpha} \langle y, w \rangle$$

for all $w \in \mathsf{E}^n$, since $\operatorname{dc} M_\alpha$ is on or above H. But since M_α is convex, this implies (see Exercise 24.1) that $y_0 \in M_\alpha$. Since this holds for each α, we have $y_0 \in \cap M_\alpha$, a contradiction. We conclude that $\operatorname{dc}(\cap M_\alpha) \subset \operatorname{cl conv}(\cup \operatorname{dc} M_\alpha)$.

To prove the reverse inclusion we note that $\cap M_\alpha \subset M_\alpha$ for each $\alpha \in \mathcal{C}$. Thus by Theorem 24.3(b), $\operatorname{dc} M_\alpha \subset \operatorname{dc}(\cap M_\alpha)$ for each $\alpha \in \mathcal{C}$, and so $\cup \operatorname{dc} M_\alpha \subset \operatorname{dc}(\cap M_\alpha)$. Since $\operatorname{dc}(\cap M_\alpha)$ is closed and convex we have $\operatorname{cl conv}(\cup \operatorname{dc} M_\alpha) \subset \operatorname{dc}(\cap M_\alpha)$. ∎

24.7. Example. For each $k = 1, 2, \ldots$, let M_k be the subset of E^1 given by $M_k \equiv \{x: x \geq k\}$. Then it is easy to see that

$$\operatorname{cl conv}\left(\bigcup_{k=1}^{\infty} \operatorname{dc} M_k\right) = \{(\alpha, x): x \leq 0\}.$$

DUAL CONES

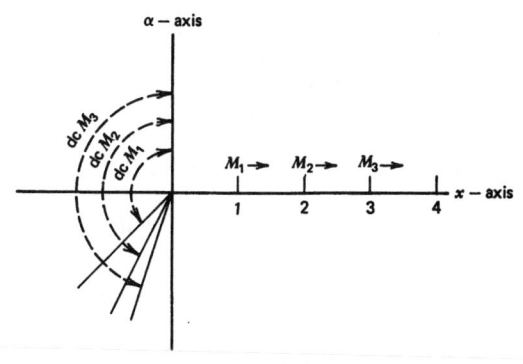

Figure 24.2.

(See Figure 24.2 and note that the α-axis is vertical.) On the other hand, $\bigcap_{k=1}^{\infty} M_k = \emptyset$, so $\mathrm{dc}(\bigcap_{k=1}^{\infty} M_k) = \mathsf{R} \times \mathsf{E}^1$. Thus we see that in Theorem 24.6 it is necessary to have at least one of the M_α compact.

The following theorem is a key result relating the intersection of convex sets with their dual cones. It will, for example, be useful in deriving the dual of the separation theorem (Exercise 24.3). It will also be essential in our duality proof of Helly's Theorem 6.3. This proof will show that Helly's Theorem can be viewed as a dual of Caratheodory's Theorem. We will conclude this section with a duality proof of the finite form of Kirchberger's Theorem 7.1. By so doing we will show in fact that it can be viewed as a dual to Helly's Theorem.

While the implications of Theorem 24.8 are far-reaching, its proof follows easily from what we have done already.

24.8. Theorem. Let M_α ($\alpha \in \mathcal{C}$) be a family of closed convex subsets of E^n. Suppose at least one of the M_α is compact. Then the following are equivalent:
  (a) $\bigcap_{\alpha \in \mathcal{C}} M_\alpha \neq \emptyset$.
  (b) The dual cones $\mathrm{dc}\, M_\alpha$ ($\alpha \in \mathcal{C}$) all have a common hyperplane of support at the origin $(0, \theta)$ in $\mathsf{R} \times \mathsf{E}^n$.
  (c) $\mathrm{conv}(\bigcup_{\alpha \in \mathcal{C}} \mathrm{int}\, \mathrm{dc}\, M_\alpha) \neq \mathsf{R} \times \mathsf{E}^n$.
  (d) $\mathrm{cl}\, \mathrm{conv}(\bigcup_{\alpha \in \mathcal{C}} \mathrm{dc}\, M_\alpha) \neq \mathsf{R} \times \mathsf{E}^n$.

PROOF. (a) $\Rightarrow$ (b) Since $\bigcap_{\alpha \in \mathcal{C}} M_\alpha$ is nonempty, its dual cone has a hyperplane of support H through $(0, \theta)$. But by Theorem 24.6, H is also a hyperplane of support for each $\mathrm{dc}\, M_\alpha$ ($\alpha \in \mathcal{C}$).

(b) $\Rightarrow$ (c) Since one of the M_α is compact, the common hyperplane of support H cannot be vertical. Hence all the dual cones are above H and thus $\mathrm{conv}(\bigcup_{\alpha \in \mathcal{C}} \mathrm{int}\, \mathrm{dc}\, M_\alpha)$ is contained in a closed half-space.

(c) ⇒ (d) If $\text{conv}(\bigcup_{\alpha \in \mathcal{C}} \text{int dc } M_\alpha)$ is not equal to $\mathsf{R} \times \mathsf{E}^n$, then it is contained in a closed half-space. Whence $\text{cl conv}(\bigcup_{\alpha \in \mathcal{C}} \text{dc } M_\alpha)$ is also contained in this same closed half-space.

(d) ⇒ (a) This follows from Theorem 24.3(c) and Theorem 24.6. ∎

24.9. Theorem. (See Helly's Theorem 6.3.) Let $\mathcal{F} \equiv \{B_\alpha : \alpha \in \mathcal{C}\}$ be a family of compact convex subsets of E^n containing at least $n + 1$ members. If every $n + 1$ members of $\mathcal{F}$ have a point in common, then all the members of $\mathcal{F}$ have a point in common.

PROOF. Suppose every $n + 1$ members of $\mathcal{F}$ have a point in common, but all the members of $\mathcal{F}$ do not. Theorem 24.8 implies that $\text{conv}(\bigcup_{\alpha \in \mathcal{C}} \text{int dc } B_\alpha) = \mathsf{R} \times \mathsf{E}^n$. Since $u \equiv (1, \theta) \in \text{int dc } B_\alpha$ for all $\alpha \in \mathcal{C}$, a variation of Caratheodory's Theorem (Exercise 2.34) implies that $(0, \theta)$ is contained in a simplex with vertices $u, x_1, x_2, \ldots, x_r$ ($r \leq n + 1$), where $x_i \in \bigcup_{\alpha \in \mathcal{C}} \text{int dc } B_\alpha$ for $i = 1, \ldots, r$. Now for each x_i ($i = 1, \ldots, r$) choose a dual cone $\text{dc } B_i$ such that $x_i \in \text{int dc } B_i$. We then have $(0, \theta) \in \text{conv}\{u, x_1, \ldots, x_r\}$ with $u \in \text{int dc } B_i$ and $x_i \in \text{int dc } B_i$ for each $i = 1, \ldots, r$. It follows that each hyperplane through $(0, \theta)$ must intersect at least one of the open cones $\text{int dc } B_i$ ($i = 1, \ldots, r$). But then Theorem 24.8 implies that $\bigcap_{i=1}^r B_i = \emptyset$. Since $r \leq n + 1$, this contradicts our original assumption. ∎

24.10. Theorem. (See Kirchberger's Theorem 7.1.) Let P and Q be nonempty finite subsets of E^n. Then P and Q can be strictly separated by a hyperplane iff for each subset T of $n + 2$ or fewer points of $P \cup Q$, there exists a hyperplane that strictly separates $T \cap P$ and $T \cap Q$.

PROOF. For each $p \in P$ and $q \in Q$, define the open half-spaces

$$F_p \equiv \{(\alpha, x): (\alpha, x) \cdot (-1, p) < 0\}$$

$$G_q \equiv \{(\alpha, x): (\alpha, x) \cdot (-1, q) > 0\}.$$

Note that $F_p = \text{int dc}\{p\}$ and G_q is the complement in $\mathsf{R} \times \mathsf{E}^n$ of $\text{dc}\{q\}$. Now let T be a subset of $n + 2$ or fewer points of $P \cup Q$. By assumption, there exists a hyperplane

$$H \equiv \{y \in \mathsf{E}^n : \langle y, u \rangle = \beta\}$$

in E^n such that

$$\langle p, u \rangle < \beta \quad \text{for all } p \in T \cap P$$

$$\langle q, u \rangle > \beta \quad \text{for all } q \in T \cap Q.$$

It follows that $(\beta, u) \in F_p \cap G_q$ for all $p \in T \cap P$ and $q \in T \cap Q$. Indeed, for

$p \in T \cap P$ and $q \in T \cap Q$ we have

$$(\beta, u) \cdot (-1, p) = \langle p, u \rangle - \beta < 0$$

and

$$(\beta, u) \cdot (-1, q) = \langle q, u \rangle - \beta > 0.$$

Thus given any $n + 2$ (or fewer) of the half-spaces F_p and G_q in $\mathsf{R} \times \mathsf{E}^n$, there exists a point in common to them. By Helly's Theorem 6.2, there exists a point, say (δ, z), in common to all the half-spaces F_p and G_q, where $p \in P$ and and $q \in Q$. It follows that the hyperplane

$$H' \equiv \{y \in \mathsf{E}^n : \langle y, z \rangle = \delta\}$$

strictly separates P and Q in E^n.

The converse is immediate. ∎

EXERCISES

24.1. Let M be a closed convex subset of E^n and let $x \in \mathsf{E}^n$. Suppose that $\langle x, w \rangle \leq \sup_{y \in M} \langle y, w \rangle$ for all $w \in \mathsf{E}^n$. Prove that $x \in M$.

24.2. Prove the following: If P is a polytope, then dc P is a polyhedral set.

24.3. Prove the following dual of the separation theorem: Let M_1 and M_2 be nonempty compact convex subsets of E^n. Then M_1 and M_2 can be strictly separated by a hyperplane iff there exists two points $p_i \in$ int dc M_i ($i = 1, 2$) such that $(0, \theta) \in$ relint $\overline{p_1 p_2}$.

*24.4. Use duality to prove Kirchberger's Theorem 7.1 for arbitrary compact subsets of E^n.

24.5. Complete the proof of Theorem 24.3.

10
OPTIMIZATION

Much of the renewed interest over the last thirty years in convex sets stems from their applications in various optimization problems. We examine two of these important applications in this chapter. Since it would require several volumes to cover these topics completely, we content ourselves with developing the basic theory and giving special attention to the role of convex sets.

In Section 25 we look at finite matrix games. Although the concepts involved can be illustrated very simply (see Example 25.1), the applications of this theory extend widely from economic decision making to military strategy.

Section 26 deals with the theory of linear programming. Once again the applications are many and varied—from nutrition to transportation to manufacturing. Virtually any company that is trying to obtain a maximum return from limited resources is dealing with a potential linear programming problem.

Both Sections 25 and 26 present graphical methods for solving simple problems. Unfortunately, these techniques are of little value in more complicated settings. In Section 27 we describe a procedure for solving linear programming problems (and matrix games) that is not restricted by the number of variables (or strategies). This method, called the simplex method, was developed by George Dantzig in the late 1940s and is now widely used in business and industry.

SECTION 25. FINITE MATRIX GAMES

In order to get a feeling for the concepts involved in the theory of games, we begin by looking at two simple examples. As in all the games considered in this section, we have two players, denoted by P_I and P_{II}, who are competing according to a fixed set of rules.

25.1. Example. Each player has a supply of pennies, nickels and dimes. At a given signal, both players display one coin. If the displayed coins are not the same, then the player showing the larger coin gets to keep both. If they are

FINITE MATRIX GAMES

both pennies or nickels, then P_{II} keeps both, but if they are both dimes then P_I keeps them. This information can be summarized by the following payoff matrix:

	p	n	d
p	−1	−1	−1
n	1	−5	−5
d	1	5	10

where the rows represent the possible plays of P_I, the columns represent the possible plays of P_{II}, and each entry indicates what P_I wins (or loses) for that particular combination of moves.

If the game is repeated several times, how might we expect the play of the game to proceed? Player I, noting that each entry in row three is positive, decides to play a dime. No matter what P_{II} may do, player I is thereby assured of at least winning a penny. Player II, on the other hand, decides to play a penny. He realizes that he cannot be certain of winning anything, but at least by playing a penny he can minimize his losses.

From a mathematical point of view, what has each player done? Player I has found the minimum of each row (the worst that could happen for that play) and chosen the row for which this minimum is largest. Similarly, player II has found the maximum of each column (the worst that can happen to him for that play) and chosen the column for which this maximum is smallest. Letting a_{ij} represent the entry in the ith row and jth column of the payoff matrix we see that

$$\max_i \min_j a_{ij} = \min_j \max_i a_{ij} = 1.$$

The entry a_{31} is called a **saddle point** (or equilibrium point) for the payoff matrix. It has the property that it is the smallest entry in its row and the largest entry in its column. (There may be other entries in its row or column of the same size.)

Thus we see that as long as both players continue to seek their best advantage, player I will always play a dime and player II will always play a penny. The situation is not quite so simple in the next example.

25.2. Example. Again we suppose that each player has a supply of pennies, nickels and dimes to play, but this time the payoff matrix is given as follows:

	p	n	d
p	10	5	−5
n	1	−1	1
d	0	−5	−10

If player I reasons as he did in the first example, he will choose to play a nickel, thereby maximizing his minimum gain (in this case a loss, i.e., -1). Player II on the other hand will select a dime. The maximum in each column (10, 5, and 1) is smallest in column 3.

Thus as the game begins, P_I continues to play a nickel and P_{II} plays a dime. After a while, however, P_{II} begins to reason: If P_I is going to play a nickel, then I'll play a nickel too so that I can win a penny. But when P_{II} starts to play his nickel over and over, then P_I begins to reason: If P_{II} is going to play a nickel then I'll play a penny so that I can win a nickel. Having done this, P_{II} then switches to a dime (to win a nickel) and then P_I starts playing a nickel... and so on. It seems that neither player can develop a winning strategy.

Mathematically speaking, we see that the payoff matrix for this game does not have a saddle point. Indeed,

$$\max_i \min_j a_{ij} = -1,$$

while

$$\min_j \max_i a_{ij} = 1.$$

This means that neither player can consistently play the same coin and be assured of optimizing his winnings. This raises the question of whether a player can formulate some combination of plays that over the long run will give him an optimal return. It is clear that such a strategy could not follow a fixed pattern, for if it did then the other player would eventually catch on and could counter each expected move. The only possibility is for the player to make each move at random, deciding only with what probability he will make each possible choice. It is, of course, not at all clear that even a complicated strategy of this sort can produce a maximal return. To see that such an optimal strategy does exist, we begin with some definitions in a more general setting.

FINITE MATRIX GAMES 157

It is clear that both the games considered in the examples are determined completely by their payoff matrices. We thus identify such a game with its matrix and speak of a **matrix game**.

25.3. Definition. Suppose that players P_I and P_{II} have, respectively, m and n possible moves or choices at each play of a matrix game. If a player selects the same choice at each play of the game, his strategy is called a **pure strategy**. A **mixed strategy** for a player is a probability distribution on the set of his pure strategies.

We may think of a mixed strategy for P_I as an m-dimensional vector $x \equiv (x_1, \ldots, x_m)$ where

$$x_i \geq 0 \quad i = 1, \ldots, m,$$

$$\sum_{i=1}^{m} x_i = 1$$

and each x_i represents the probability with which P_I selects the ith possible move. Similarly, we may think of a mixed strategy for P_{II} as an n-dimensional vector $y \equiv (y_1, \ldots, y_n)$ with corresponding restrictions. In this point of view, each pure strategy for P_I corresponds to one of the standard basis vectors $e_1 \equiv (1, 0, \ldots, 0), \ldots, e_m \equiv (0, \ldots, 0, 1)$ in E^m, and a mixed strategy is a point in their convex hull. This is precisely what enables us to apply the theory of convex sets to the study of matrix games. If we let X denote the set of all (mixed) strategies for P_I and Y the set of all (mixed) strategies for P_{II}, then X is an $(m-1)$-simplex in E^m and Y is an $(n-1)$-simplex in E^n.

Suppose now that P_I and P_{II} are playing the matrix game $A \equiv (a_{ij})$, where a_{ij} is, as usual, the entry in the ith row and the jth column of A. If P_I uses the strategy $x \equiv (x_1, \ldots, x_m)$ and P_{II} uses the strategy $y \equiv (y_1, \ldots, y_n)$, then it follows from elementary probability theory that the **expected payoff** to P_I will be

$$E(x, y) \equiv \sum_{i=1}^{m} \sum_{j=1}^{n} x_i a_{ij} y_j.$$

In matrix notation we have $E(x, y) = \langle x, Ay \rangle = \langle xA, y \rangle = xAy$, where y is written as a column vector.

Suppose now that P_I were to choose a particular strategy, say $\bar{x}$. If P_{II} discovers this strategy, then P_{II} would certainly choose y so as to minimize

$$E(\bar{x}, y) = \sum_i \sum_j \bar{x}_i a_{ij} y_j.$$

Since $E(\bar{x}, y)$ is linear in y (Exercise 25.6), it attains its minimum at one of the extreme points of Y (Theorem 5.7), that is, when y is one of the pure strategies.

Thus the value $v(x)$ of a particular strategy x to P_I is given by

$$v(x) \equiv \min_{y \in Y} E(x, y) = \min_{j} \sum_{i} x_i a_{ij}.$$

If P_I chooses x so as to maximize its value, then we have

$$v_I \equiv \max_{x \in X} v(x) = \max_{x \in X} \min_{j} \sum_{i} x_i a_{ij}.$$

A similar analysis for P_{II} shows that a particular strategy y will have a value $v(y)$ given by

$$v(y) \equiv \max_{x \in X} E(x, y) = \max_{i} \sum_{j} a_{ij} y_j$$

so that

$$v_{II} \equiv \min_{y \in Y} v(y) = \min_{y \in Y} \max_{i} \sum_{j} a_{ij} y_j.$$

The two numbers v_I and v_{II} are called the **values** of the game to P_I and P_{II}, respectively. The value v_I can be thought of as the most that P_I can guarantee himself independent of what P_{II} may do. Similarly, v_{II} is the least that P_{II} will have to lose regardless of what P_I may do. It is easy to see that $v_I \leq v_{II}$:

25.4. Theorem. In any matrix game, $v_I \leq v_{II}$.

PROOF. Since $v(x) = \min_{y \in Y} E(x, y)$ and $v(y) = \max_{x \in X} E(x, y)$, we have

$$v(x) \leq E(x, y) \leq v(y)$$

for all $x \in X$ and $y \in Y$. Thus

$$\max_{x \in X} v(x) \leq \min_{y \in Y} v(y),$$

which is what we wanted to show. ∎

Before proceeding further with the theory, let us return to Example 25.2. Recall that the payoff matrix was given by

$$M = \begin{bmatrix} 10 & 5 & -5 \\ 1 & -1 & 1 \\ 0 & -5 & -10 \end{bmatrix}.$$

FINITE MATRIX GAMES

Suppose we let $x = (\frac{1}{2}, \frac{1}{4}, \frac{1}{4})$ and $y = (\frac{1}{4}, \frac{1}{2}, \frac{1}{4})$. Then

$$E(x, y) = \frac{10}{8} + \frac{5}{4} - \frac{5}{8}$$
$$+ \frac{1}{16} - \frac{1}{8} + \frac{1}{16}$$
$$+ 0 - \frac{5}{8} - \frac{10}{16}$$
$$= \frac{5}{8}$$

Furthermore,

$$E(x, e_1) = \frac{10}{2} + \frac{1}{4} + 0 = \frac{21}{4},$$
$$E(x, e_2) = \frac{5}{2} - \frac{1}{4} - \frac{5}{4} = 1,$$
$$E(x, e_3) = -\frac{5}{2} + \frac{1}{4} - \frac{10}{4} = -\frac{19}{4},$$

so that $v(x) = \min\{\frac{21}{4}, 1, -\frac{19}{4}\} = -\frac{19}{4}$. Similarly, we find that $v(y) = \max\{\frac{15}{4}, 0, -\frac{10}{2}\} = \frac{15}{4}$. Notice that $v(x) \leq E(x, y) \leq v(y)$, as we would expect.

In the preceding theorem it was easy to show that $v_I \leq v_{II}$. One of the remarkable results in game theory is that in fact, $v_I = v_{II}$.

25.5. Theorem (Minimax Theorem). In any matrix game, $v_I = v_{II}$. That is,

$$\max_{x \in X} \min_{y \in Y} E(x, y) = \min_{y \in Y} \max_{x \in X} E(x, y).$$

Before proving the Minimax Theorem, let us comment first on some of its consequences. The following definition is standard.

25.6. Definition. Using the preceding notation, we let $v = v_I = v_{II}$ denote the **value** of the game. The strategies $\bar{x}$ and $\bar{y}$ for P_I and P_{II}, respectively, are called **optimal** if

$$v(\bar{x}) = v = v(\bar{y}).$$

Equivalently, $\bar{x}$ is optimal for P_I if

$$E(\bar{x}, y) \geq v \quad \text{for all} \quad y \in Y$$

and $\bar{y}$ is optimal for P_{II} if

$$E(x, \bar{y}) \leq v \quad \text{for all} \quad x \in X.$$

Any pair of optimal strategies $(\bar{x}, \bar{y})$ is called a **solution** to the game.

The following question now arises: Does every matrix game have a solution? That is, do optimal strategies for the two players always exist? The reassuring answer is affirmative, and this is an easy consequence of the Minimax Theorem.

25.7. Theorem (Fundamental Theorem for Matrix Games). In any matrix game, there are always optimal mixed strategies. That is, every matrix game has a solution.

PROOF. Since $v(x)$ is continuous on the compact set X, Theorem 1.21 implies that there exists a point $\bar{x}$ in X such that

$$v(\bar{x}) = \max_{x \in X} v(x) \equiv v_\mathrm{I}.$$

Similarly, there exists $\bar{y}$ in Y such that

$$v(\bar{y}) = \min_{y \in Y} v(y) \equiv v_\mathrm{II}.$$

According to the Minimax Theorem, $v_\mathrm{I} = v_\mathrm{II} = v$. ∎

We now turn our attention to proving the Minimax Theorem. Many proofs are known, the first having been given by J. von Neumann in 1928. As is frequently the case, this first proof was quite complicated, but more elementary techniques have since been found. The argument presented here is basically the same as that given by von Neumann and Morgenstern in their classical book, *Theory of Games and Economic Behavior* (1944). It is based on separating a point from a convex set by a hyperplane. We begin with two lemmas.

25.8. Lemma (Theorem of the Supporting Hyperplanes). Let B be a closed convex subset of E^m. If $\theta \notin B$, then there exist real numbers $s_1, \ldots, s_m$ such that

$$b_1 s_1 + \cdots + b_m s_m > 0$$

for each point $b \equiv (b_1, \ldots, b_m)$ in B.

PROOF. Since $\theta \notin B$, Theorem 4.12 implies that there exists a hyperplane $H \equiv [f: \alpha]$ which strictly separates θ and B, and we may assume without loss of generality that $f(B) > \alpha > 0$. By Corollary 3.5 there exists a point $u \equiv (s_1, \ldots, s_m)$ in E^m such that $f(x) = \langle x, u \rangle$ for all x in E^m. Thus for all $b \equiv (b_1, \ldots, b_m)$ in B we have

$$b_1 s_1 + \cdots + b_m s_m = \langle b, u \rangle = f(b) > \alpha > 0. \quad \blacksquare$$

25.9. Lemma (Theorem of the Alternative for Matrices). Let $A \equiv (a_{ij})$ be an $m \times n$ matrix. Then one of the following two cases must hold:

FINITE MATRIX GAMES   **161**

CASE 1. The origin θ in E^m is contained in $B \equiv \mathrm{conv}\{a_1, \ldots, a_n, e_1, \ldots, e_m\}$ where $a_1 \equiv (a_{11}, \ldots, a_{m1}), \ldots, a_n \equiv (a_{1n}, \ldots, a_{mn})$ are the column vectors of A and $e_1 \equiv (1, 0, \ldots, 0), \ldots, e_m \equiv (0, \ldots, 0, 1)$ are the standard basis for E^m.

CASE 2. There exist real numbers $x_1, \ldots, x_m$ such that

$$x_i > 0, \quad i = 1, \ldots, m$$

$$\sum_{i=1}^{m} x_i = 1$$

and

$$\sum_{i=1}^{m} x_i a_{ij} > 0 \quad \text{for } j = 1, \ldots, n.$$

PROOF. Suppose Case 1 does not hold. Then by Lemma 25.8 there exist real numbers $s_1, \ldots, s_m$ such that

$$\sum_{i=1}^{m} b_i s_i > 0$$

for all $b \equiv (b_1, \ldots, b_m)$ in B. In particular this will hold if b is one of the $m + n$ vectors a_i or e_j. Thus

$$\sum_{i=1}^{m} a_{ij} s_i > 0 \quad \text{for all } j$$

and

$$s_i > 0 \quad \text{for all } i.$$

It follows that $\sum_{i=1}^{m} s_i > 0$ and we can define

$$x_i \equiv \frac{s_i}{\sum s_i} \quad i = 1, \ldots, m$$

and this clearly satisfies the required conditions. ∎

PROOF OF THEOREM 25.5. Let A be a matrix game. By Lemma 25.9, either Case 1 or Case 2 must hold.

If Case 1 holds, then θ is a convex combination of the $m + n$ points $a_1, \ldots, a_n, e_1, \ldots, e_m$. Thus there exist real numbers $t_1, \ldots, t_{m+n}$ such that

$$t_k \geq 0, \quad k = 1, \ldots, m + n$$

$$\sum_{k=1}^{m+n} t_k = 1$$

and

$$\sum_{k=1}^{n} t_k a_k + \sum_{k=n+1}^{m+n} t_k e_{k-n} = \theta.$$

Thus for each coordinate $i = 1, \ldots, m$ we have

$$\sum_{j=1}^{n} t_j a_{ij} + t_{n+i} = 0.$$

Now the sum $t_1 + \cdots + t_n$ must be positive, for otherwise $t_1, \ldots, t_n$ would all be zero (and hence so would $t_{n+1}, \ldots, t_{m+n}$) and the sum of all the t_k would not be equal to 1. Thus we can let

$$y_j = \frac{t_j}{\sum_{j=1}^{n} t_j} \qquad j = 1, \ldots, n$$

and we have

$$y_j \geq 0, \qquad \text{for all } j = 1, \ldots, n,$$

$$\sum_{j=1}^{n} y_j = 1,$$

and

$$\sum_{j=1}^{n} a_{ij} y_j = \frac{-t_{n+i}}{\sum_{j=1}^{n} t_j} \leq 0 \qquad \text{for all } i = 1, \ldots, m.$$

It follows that

$$v(\tilde{y}) \equiv \max_{i=1,\ldots,m} \sum_{j=1}^{n} a_{ij} y_j \leq 0$$

for this particular $\tilde{y} \equiv (y_1, \ldots, y_n)$, and so also

$$v_{\text{II}} \equiv \min_{y \in Y} v(y) \leq 0.$$

Now suppose that Case 2 holds. Then for $\tilde{x} \equiv (x_1, \ldots, x_m)$ we have

$$v(\tilde{x}) \equiv \min_{j=1,\ldots,n} \sum_{i=1}^{m} a_{ij} x_i > 0$$

and so $v_{\text{I}} \equiv \max_{x \in X} v(x) > 0$.

FINITE MATRIX GAMES

We have shown that it is not possible to have

$$v_{\mathrm{I}} \leq 0 < v_{\mathrm{II}}.$$

Let us now consider another matrix $B \equiv (b_{ij})$, where

$$b_{ij} = a_{ij} - \alpha$$

for an arbitrary constant α. Since for any $x \equiv (x_1, \ldots, x_m)$ in X and $y \equiv (y_1, \ldots, y_n)$ in Y we have $\Sigma x_i = 1$ and $\Sigma y_j = 1$ it follows that

$$\sum_{i=1}^{m} \sum_{j=1}^{n} x_i(a_{ij} - \alpha)y_j = \sum_{i=1}^{m} \sum_{j=1}^{n} x_i a_{ij} y_j - \alpha.$$

Thus

$$v_{\mathrm{I}}(B) = v_{\mathrm{I}}(A) - \alpha$$

and

$$v_{\mathrm{II}}(B) = v_{\mathrm{II}}(A) - \alpha.$$

The first part of this proof applied to B implies that it is not possible to have

$$v_{\mathrm{I}}(B) \leq 0 < v_{\mathrm{II}}(B),$$

so that it is also not possible to have

$$v_{\mathrm{I}}(A) \leq \alpha < v_{\mathrm{II}}(A).$$

Since α was arbitrary, this means we cannot have $v_{\mathrm{I}} < v_{\mathrm{II}}$. But Theorem 25.4 says that $v_{\mathrm{I}} \leq v_{\mathrm{II}}$, so in fact $v_{\mathrm{I}} = v_{\mathrm{II}}$. ∎

We have now proved that each matrix game has a solution, that is, that optimal strategies exist for both players. We have said nothing, however, about how these optimal strategies may be found, and the proof of our existence theorem gives us no hint. To remedy this situation we begin by showing how a large matrix game may sometimes be reduced to a smaller one, and then finally we give a geometric method of solving any $2 \times n$ matrix game.

25.10. Definition. An n-tuple $a \equiv (a_1, \ldots, a_n)$ is said to **dominate** a second n-tuple $b \equiv (b_1, \ldots, b_n)$ if

$$a_i \geq b_i \quad \text{for all } i$$

and

$$a_i > b_i \quad \text{for at least one } i.$$

Suppose that in the matrix game A, row i dominates row j. This means that the pure strategy i for P_I is at least as good as the pure strategy j, no matter what P_{II} may choose, and for some choice of P_{II}, the pure strategy i for P_I is better. It follows that pure strategy j (the "smaller" one) can be ignored by P_I without hurting his expected payoff. A similar analysis applies to the columns of A, where in this case the dominating (i.e., "larger") column is ignored. These observations are formalized in the following theorem.

25.11. Theorem. Let A be an $m \times n$ matrix game. If row j in the matrix A *is dominated by* some other row, then let B be the $(m-1) \times n$ matrix obtained by deleting row j from A. Similarly, if column k of matrix A *dominates* some other column, let C be the $m \times (n-1)$ matrix obtained by deleting column k from A. In either case, any optimal strategy of the reduced matrix game B or C will also be an optimal strategy for A.

25.12. Example. The process described may, of course, be repeated. Consider the game with matrix

$$A \equiv \begin{bmatrix} 5 & 3 & 1 & 0 \\ 6 & 1 & 4 & 5 \\ 4 & 5 & 3 & 3 \end{bmatrix}.$$

Since the first column dominates the third, P_{II} will never want to use his first pure strategy. Deleting column 1, we obtain

$$\begin{bmatrix} 3 & 1 & 0 \\ 1 & 4 & 5 \\ 5 & 3 & 3 \end{bmatrix}.$$

In this matrix, row 1 is dominated by row 3. Deleting row 1 we have

$$\begin{bmatrix} 1 & 4 & 5 \\ 5 & 3 & 3 \end{bmatrix},$$

which can be reduced further by dropping the last column, since it dominates column 2. Thus the original matrix game has been reduced to

$$B \equiv \begin{bmatrix} 1 & 4 \\ 5 & 3 \end{bmatrix}$$

and any optimal strategy for B will also be optimal for A. (Remember that the first row of B corresponds to the second row of A, the second column of B corresponds to the third column of A, etc.)

We now turn our attention to a $2 \times n$ matrix game, where P_I has two

FINITE MATRIX GAMES

possible pure strategies and P_{II} has n. Thus

$$A \equiv \begin{bmatrix} a_{11} & a_{12} & \cdots & a_{1n} \\ a_{21} & a_{22} & \cdots & a_{2n} \end{bmatrix}.$$

The objective of P_I is to maximize

$$v(x) = \min_j \{a_{1j}x_1 + a_{2j}x_2\}.$$

Since $x_2 = 1 - x_1$, this is equivalent to

$$v(x) = \min_j \{(a_{1j} - a_{2j})x_1 + a_{2j}\}.$$

Thus $v(x)$ is the minimum of n linear functions of the single variable x_1. If we graph each of these functions, then it will be easy to maximize their minimum.

25.13. Example. Consider the game with matrix

$$\begin{bmatrix} 4 & 0 & 1 & 2 \\ 1 & 5 & 3 & 5 \end{bmatrix}$$

The lines $f_j(x_1) = (a_{1j} - a_{2j})x_1 + a_{2j}$ are easy to graph since they go through the points $(0, a_{2j})$ and $(1, a_{1j})$.

In Figure 25.1, the heavy bent line at the bottom represents the function $v(x)$. The highest point on this line, M, is at the intersection of the lines corresponding to the first and third columns of the matrix. The coordinates of M are found to be $(\frac{2}{5}, \frac{11}{5})$. The first coordinate $\frac{2}{5}$ gives the optimal probability for the first pure strategy. Thus $\bar{x} = (\frac{2}{5}, \frac{3}{5})$ is the optimal mixed strategy for P_I. The second coordinate of M is the maximum of $v(x)$ (i.e., the value of the game). Thus $v = \frac{11}{5}$.

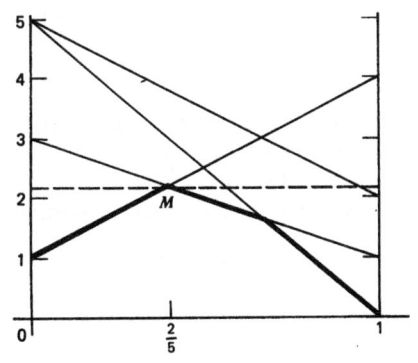

Figure 25.1.

(a)   (b)   (c)

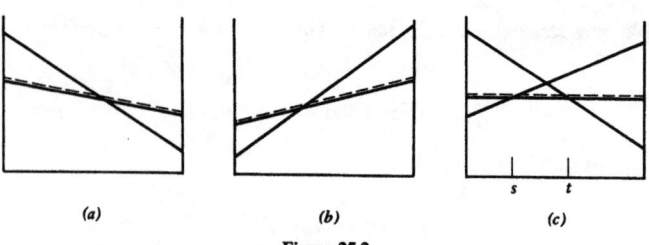

Figure 25.2.

The optimal strategy for P_{II} will be a combination of his <u>first</u> and <u>third</u> pure strategies. Graphically, he wants to have a strategy corresponding to the horizontal line through M. By so doing, he can guarantee that his losses will be no greater than the value of the game (which is the best he can hope to do). Now the horizontal line through M is three-fifths of the way from line 1 to line 3. Since it is closer to line 3, y_3 will be weighted heavier, and his optimal mixed strategy will be $\bar{y} = (\frac{2}{5}, 0, \frac{3}{5}, 0)$. Notice that the fourth column in the matrix dominates the second, corresponding to its graph lying above the graph of the second. Thus the fourth pure strategy could have been eliminated from the beginning.

There are, of course, other possible patterns to the strategy lines. They are illustrated in Figure 25.2, where the solid lines represent P_{II}'s pure strategies and the broken line is his optimal one (which ends up being a pure strategy in each case). In Figure 25.2a, $\bar{x} = (1, 0)$. In Figure 25.2b, $\bar{x} = (0, 1)$. In Figure 25.2c, $\bar{x} = (s, 1 - s)$ or $(t, 1 - t)$ or any strategy in between them.

EXERCISES

25.1. Let M be the matrix game having payoff matrix

$$\begin{bmatrix} 2 & 0 & 1 & -1 \\ -1 & 1 & -2 & 0 \\ 1 & -2 & 2 & 1 \end{bmatrix}.$$

Find $E(x, y)$, $v(x)$ and $v(y)$ when x and y have the given values.
(a) $x = (\frac{1}{3}, 0, \frac{2}{3})$ and $y = (\frac{1}{4}, \frac{1}{2}, 0, \frac{1}{4})$
(b) $x = (\frac{1}{4}, \frac{1}{2}, \frac{1}{4})$ and $y = (0, \frac{1}{2}, \frac{1}{4}, \frac{1}{4})$

25.2. Let M be the matrix game having payoff matrix

$$\begin{bmatrix} 1 & 2 & -2 \\ 0 & 1 & 3 \\ 4 & -1 & 1 \end{bmatrix}.$$

Let $x = y = (\frac{1}{3}, \frac{1}{2}, \frac{1}{6})$. Find $v(x)$, $v(y)$ and v.

EXERCISES

25.3. Find optimal strategies and the value of each matrix game.

(a) $\begin{bmatrix} 1 & 3 \\ 4 & 5 \end{bmatrix}$ (b) $\begin{bmatrix} 1 & -3 \\ -4 & 5 \end{bmatrix}$

(c) $\begin{bmatrix} 1 & 0 \\ 2 & 2 \end{bmatrix}$ (d) $\begin{bmatrix} 3 & 2 & 5 & 2 \\ -1 & 9 & 0 & 8 \end{bmatrix}$

(e) $\begin{bmatrix} 4 & 5 & 2 & 0 \\ 1 & 3 & 2 & 5 \end{bmatrix}$ (f) $\begin{bmatrix} 3 & -1 & 0 \\ 2 & 0 & 3 \\ -1 & -2 & 1 \end{bmatrix}$

(g) $\begin{bmatrix} 0 & 1 & -1 & 2 & 3 \\ 0 & -1 & 2 & -1 & -2 \\ 2 & -1 & 4 & 0 & -2 \\ 0 & 0 & -2 & 1 & 2 \end{bmatrix}$ (h) $\begin{bmatrix} 7 & 0 & 6 & 1 \\ 4 & 4 & 3 & 4 \\ 6 & 2 & 5 & 3 \\ 4 & 7 & 2 & 8 \end{bmatrix}$

25.4. Find the optimal strategies and the value of the game in Example 25.2.

25.5. Consider the matrix game $A \equiv \begin{bmatrix} a & b \\ c & d \end{bmatrix}$. Suppose neither row (or column) dominates the other.
  (a) Find a formula for the optimal strategy for P_I.
  (b) Let $J \equiv \begin{bmatrix} 1 & 1 \\ 1 & 1 \end{bmatrix}$ and let α and β be real numbers with $\alpha \neq 0$. Use your answer to part (a) to show that the optimal strategy for P_I in the matrix game $B \equiv \alpha A + \beta J$ is the same as in A. In particular, note that the optimal strategies in A and $-A$ are the same.

25.6. Let $\bar{x}$ be a fixed strategy for P_I. Prove that $E(\bar{x}, y)$ is a linear function for $y \in Y$.

25.7. A certain army is engaged in guerrilla warfare. It has two routes to get supplies to its troops: the river road or overland through the jungle. On a given day, the guerrillas can attack only one of the two roads. If the convoy goes along the river and the guerrillas attack, the convoy will have to turn back, and six soldiers will be lost. If the convoy goes overland and the guerrillas attack, half the convoy will get through, but eight soldiers will be lost. Each day a supply convoy travels one of the roads, and if the guerrillas are watching the other road, it gets through with no losses.
  (a) What is the optimal strategy for the army if it wants to maximize the amount of supplies it gets to its troops?
  (b) What is the optimal strategy for the army if it wants to minimize its casualties?

(c) What is the optimal strategy for the guerrillas if they want to prevent the most supplies from getting through?

(d) What is the optimal strategy for the guerrillas if they want to inflict maximum losses on the army?

(e) Suppose that whenever the convoy goes overland two soldiers are lost due to land mines (whether they are attacked or not). Find the optimal strategies for the army and the guerrillas with respect to the number of army casualties.

25.8. Let A be a matrix game having value v. Find an example to show that $E(x, y) = v$ does not necessarily imply that x and y are optimal strategies.

25.9. Suppose that the matrix game A has value $v = 0$. Prove that $\bar{x} \in X$ is optimal for P_I iff $\bar{x}A \geq \theta$. Note: we interpret an inequality between two vectors as applying to each of their corresponding coordinates.

25.10. A square matrix $A \equiv (a_{ij})$ is called **skew-symmetric** if $a_{ij} = -a_{ji}$ for all i and j. A matrix game is called **symmetric** if its matrix is skew-symmetric. Prove that the value of a symmetric game is zero. Moreover, prove that if x is optimal for P_I, then x is also optimal for P_{II}.

25.11. Prove that the set of optimal strategies for each player in a matrix game is a polytope.

25.12. Prove directly (without using Exercises 25.9 and 25.11) that the set of optimal strategies for each player in a matrix game is a convex set.

*25.13. Let $B \equiv \{e_1, \ldots, e_m\}$ be the standard basis for $\mathbf{E}^m$. Prove the following: If $\bar{X}$ is any polytope in conv B, then there exists a matrix game having $\bar{X}$ as the set of all optimal strategies for P_I.

SECTION 26. LINEAR PROGRAMMING

Linear programming, simply stated, involves the maximizing (or minimizing) of a linear function (called the **objective function**) subject to certain linear constraints. In many situations, the constraints take the form of linear inequalities and the variables are restricted to nonnegative values. To be more precise, we now define the so-called "canonical form" of the linear programming problem.

26.1. Definition. Given an $m \times n$ matrix $A \equiv (a_{ij})$, an m-tuple $b \equiv (b_1, \ldots, b_m)$, and an n-tuple $c \equiv (c_1, \ldots, c_n)$, the **canonical linear programming problem** is the following:

Find an n-tuple $x \equiv (x_1, \ldots, x_n)$ to maximize

$$f(x_1, \ldots, x_n) \equiv c_1 x_1 + c_2 x_2 + \cdots + c_n x_n$$

subject to the constraints

$$a_{11}x_1 + a_{12}x_2 + \cdots + a_{1n}x_n \leq b_1$$
$$a_{21}x_1 + a_{22}x_2 + \cdots + a_{2n}x_n \leq b_2$$
$$\vdots \qquad\qquad\qquad\qquad \vdots$$
$$a_{m1}x_1 + a_{m2}x_2 + \cdots + a_{mn}x_n \leq b_m$$

and

$$x_j \geq 0 \quad \text{for } j = 1, \ldots, n.$$

This may be restated in vector-matrix notation as follows:

$$\text{Maximize} \quad f(x) \equiv \langle c, x \rangle \tag{1}$$

subject to the constraints

$$Ax \leq b \tag{2}$$

and

$$x \geq \theta, \tag{3}$$

where we take x and b to be column vectors this time and where we interpret an inequality between two vectors as applying to each of their coordinates.

Any vector x which satisfies (2) and (3) is called a **feasible solution**, and the set of all feasible solutions, denoted by $\mathcal{F}$, is called the **feasible set**. A vector $\bar{x}$ in $\mathcal{F}$ is an **optimal solution** if $f(\bar{x}) = \max_{x \in \mathcal{F}} f(x)$.

The canonical statement of the problem is really not as restrictive as it might seem. If we want to minimize a function $h(x)$, we replace it with the problem of maximizing $-h(x)$. A constraint inequality of the sort

$$a_{i1}x_1 + \cdots + a_{in}x_n \geq b_i$$

can be replaced by

$$-a_{i1}x_1 - \cdots - a_{in}x_n \leq -b_i,$$

and an equality constraint

$$a_{i1}x_1 + \cdots + a_{in}x_n = b_i$$

can be replaced by two inequalities

$$a_{i1}x_1 + \cdots + a_{in}x_n \leq b_i$$
$$-a_{i1}x_1 - \cdots - a_{in}x_n \leq -b_i.$$

Unlike our experience with matrix games, it is quite possible for a linear programming problem (even a simple one) to have no optimal solution. Two things can go wrong. It may be that the constraint inequalities are inconsistent so that $\mathcal{F} = \emptyset$. In this case, we say that the program is **infeasible**. Or it may be that the objective function takes on arbitrarily large values in $\mathcal{F}$ so that the desired maximum does not exist. In this case, we say that the program is **unbounded**.

26.2. Examples. The program

$$\text{Maximize} \quad 3x$$
$$\text{subject to} \quad x \leq 2$$
$$-x \leq -3$$
$$x \geq 0$$

is infeasible, while the program

$$\text{Maximize} \quad 3x$$
$$\text{subject to} \quad -x \leq 4$$
$$x \geq 0$$

is unbounded.

Fortunately, these are the only two things that can go wrong. This is verified in Theorem 26.4, which follows an important lemma.

26.3. Lemma. The set $\mathcal{F}$ of feasible solutions to a linear programming problem is a polyhedral set containing no lines.

PROOF. Each of the constraints in (2) and (3) is a linear inequality that determines a closed half-space in $\mathbf{E}^n$. Since $\mathcal{F}$ is equal to the intersection of this finite number of closed half-spaces, $\mathcal{F}$ is polyhedral. The requirement (3) that $x \geq \theta$ means that $\mathcal{F}$ lies in the nonnegative orthant, and hence contains no lines. ∎

LINEAR PROGRAMMING

26.4. Theorem. If the objective function is bounded above on $\mathcal{F}$ and $\mathcal{F} \neq \varnothing$, then the linear programming problem has at least one optimal solution. Furthermore, at least one of the optimal solutions is an extreme point of $\mathcal{F}$.

PROOF. This follows from Lemma 26.3 and Theorem 20.10. ∎

Not only does Theorem 26.4 tell us when an optimal solution must exist, but it also suggests a possible technique for finding one: We evaluate the objective function at each of the extreme points of $\mathcal{F}$ and select the point that gives the largest value. This works fine in simple cases (see Example 26.5), but unfortunately in many practical applications it is essentially impossible. A typical linear programming problem in industry may involve hundreds of variables and constraints. Even with high-speed computers it would be exceedingly time-consuming to have to find *all* the extreme points of $\mathcal{F}$. (The simultaneous solution of each subset of m of the $m + n$ constraints yields a potential extreme point, which may or may not be feasible. Thus $\binom{m+n}{m} \equiv \dfrac{(m+n)!}{m!n!}$ possible extreme points must be checked for feasibility.)

One technique of finding an optimal solution that avoids most of these problems is the so-called "simplex method." (See Section 27.) The idea behind this method is as follows:

(a) Select an extreme point x of $\mathcal{F}$. (b) Consider all the edges that join at x. If the objective function f cannot be increased by moving along any of these edges, then x is an optimal solution. (c) If, on the other hand, f can be increased by moving along some of the edges, we choose the one that gives the largest increase and move to the extreme point at the opposite end. (d) The process is then repeated beginning at step (b). Since f is increased at each step, it follows that we cannot go through the same extreme point twice. Since there are only a finite number of extreme points, this process will bring us to an optimal solution (if there is one) in a finite (hopefully, small) number of steps. If the program is unbounded, then eventually we will come to an unbounded edge at step (c) along which f increases without bound.

We now consider a particular example that illustrates the application of Theorem 26.4 in a graphical setting.

26.5. Example

$$\text{Maximize} \quad f(x_1, x_2) \equiv 2x_1 + 3x_2$$

$$\text{subject to} \quad x_1 \leq 30$$

$$x_2 \leq 20$$

$$x_1 + 2x_2 \leq 52$$

$$\text{and } x_1 \geq 0, x_2 \geq 0.$$

172   **OPTIMIZATION**

Figure 26.1.

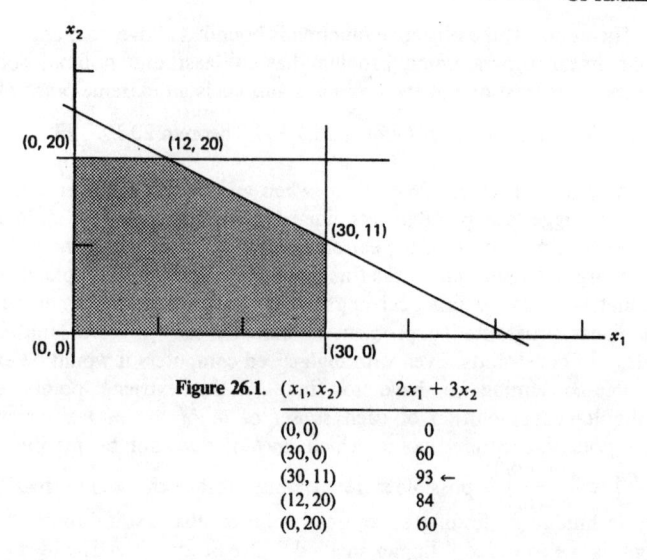

(x_1, x_2)	$2x_1 + 3x_2$
(0, 0)	0
(30, 0)	60
(30, 11)	93 ←
(12, 20)	84
(0, 20)	60

In Figure 26.1 we have graphed each of the constraint inequalities to obtain the pentagonal feasible set, which is shaded. The extreme points are found by solving the appropriate pairs of linear equations. Upon evaluating the objective function at each extreme point (see the table), we find that $x_1 = 30$, $x_2 = 11$ gives the maximum value of 93.

Another geometric technique that can be used when the problem involves two variables is to graph several level lines for the objective function. (See Figure 26.2.) It is clear that the line through $(30, 11)$ corresponds to the maximum value of $f(x)$ over the feasible set.

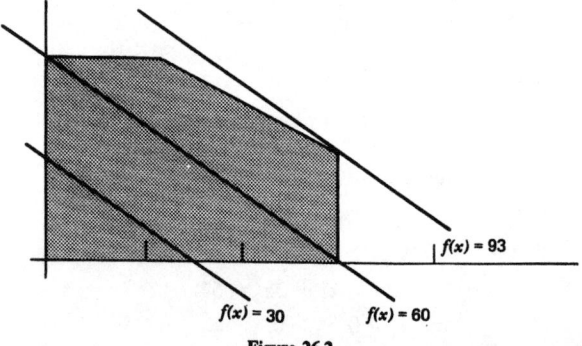

Figure 26.2.

LINEAR PROGRAMMING

One of the important aspects of linear programming is the relationship of the canonical problem, called the primal problem, to another problem, called the dual problem.

Primal Problem P	Dual Problem P^*
Maximize	Minimize
(1) $f(x) \equiv \langle c, x \rangle$	(1*) $g(y) \equiv \langle b, y \rangle$
subject to	subject to
(2) $Ax \leq b$	(2*) $yA \geq c$
(3) $x \geq \theta$	(3*) $y \geq \theta$

Notationally, it is sometimes useful to represent a vector either as a row vector or a column vector. Thus the inequality $yA \geq c$ implies that y and c are both row vectors. This is equivalent to $A^t y \geq c$ (A^t being the transpose of A), where we now interpret y and c as column vectors.

If we write the dual problem in canonical form we have:

$$\text{Maximize} \quad h(y) \equiv \langle -b, y \rangle$$

$$\text{subject to} \quad -A^t y \leq -c \text{ and } y \geq \theta.$$

Thus the dual of the dual problem is

$$\text{Minimize} \quad F(x) \equiv \langle -c, x \rangle$$

$$\text{subject to} \quad x(-A)^t \geq -b, \qquad x \geq \theta$$

or, in canonical form,

$$\text{Maximize} \quad G(x) \equiv \langle c, x \rangle$$

$$\text{subject to} \quad Ax \leq b, \qquad x \geq \theta,$$

which is the same as the primal problem. Thus the dual of the dual problem is the original primal problem.

In game theory it was easy to show (Theorem 25.4) that $v_I \leq v_{II}$ (i.e., that the value of the maximizing problem for P_I was less than or equal to the value of the minimizing problem for P_{II}). This corresponds in linear programming to the value of the primal problem [the maximum $f(x)$] being less than or equal to the value of the dual problem [the minimum $g(y)$]. This also is not difficult to prove.

26.6. Theorem. Let P be a (primal) linear programming problem with feasible set $\mathcal{F}$, and let P^* be the dual problem with feasible set $\mathcal{F}^*$.

(a) If $x \in \mathcal{F}$ and $y \in \mathcal{F}^*$, then $f(x) \leq g(y)$.

(b) If for some $\bar{x} \in \mathcal{F}$ and $\bar{y} \in \mathcal{F}^*$ we have $f(\bar{x}) = g(\bar{y})$, then $\bar{x}$ is an optimal solution to P and $\bar{y}$ is an optimal solution to P^*.

PROOF. (a) If $x \in \mathcal{F}$ and $y \in \mathcal{F}^*$, then from (2) and (2*) we obtain

$$f(x) = \langle c, x \rangle$$
$$\leq \langle yA, x \rangle$$
$$= \langle y, Ax \rangle$$
$$\leq \langle y, b \rangle = g(y).$$

(b) If $f(\bar{x}) = g(\bar{y})$, then for any $x \in \mathcal{F}$ we have

$$f(x) \leq g(\bar{y}) = f(\bar{x})$$

by part (a), so $\bar{x}$ is an optimal solution to P. Similarly, for any $y \in \mathcal{F}^*$ we have

$$g(y) \geq f(\bar{x}) = g(\bar{y}),$$

so that $\bar{y}$ is an optimal solution to P^*. ∎

We now come to the fundamental duality theorem of linear programming. After stating the theorem we derive two important lemmas that are used in the proof of the theorem. The first lemma is based on the idea of separating two disjoint convex sets by a hyperplane.

26.7. Theorem (Duality Theorem). Let P be a (primal) linear programming problem with feasible set $\mathcal{F}$, and let P^* be the dual program with feasible set $\mathcal{F}^*$.

(a) If $\mathcal{F}$ and $\mathcal{F}^*$ are both nonempty, then P and P^* both have optimal solutions, say $\bar{x}$ and $\bar{y}$, respectively. Furthermore $f(\bar{x}) = g(\bar{y})$.

(b) If either $\mathcal{F}$ or $\mathcal{F}^*$ is empty, then neither P nor P^* has an optimal solution.

(c) If one of the problems P or P^* has an optimal solution $\bar{x}$ or $\bar{y}$, respectively, then so does the other, and $f(\bar{x}) = g(\bar{y})$.

26.8. Lemma (Farkas Lemma). Exactly one of the following alternatives holds. Either the equation

$$Ax = b \tag{4}$$

has a nonnegative solution, or the inequalities

$$yA \geq \theta, \quad \langle b, y \rangle < 0 \tag{5}$$

have a solution.

LINEAR PROGRAMMING

PROOF. Suppose that (4) were to have a nonnegative solution x and (5) were to have a solution y. Then

$$0 > \langle b, y \rangle = \langle Ax, y \rangle = \langle x, yA \rangle.$$

But $x \geq \theta$ and $yA \geq \theta$, so $\langle x, yA \rangle \geq 0$, a contradiction. Thus we conclude that both alternatives cannot hold.

Now suppose that (4) has no nonnegative solution. Then the sets $\{b\}$ and

$$S \equiv \{z \in \mathsf{E}^m : z = Ax, x \geq \theta\}$$

are disjoint. It is easy to see (Exercise 26.4) that S is a closed convex cone with vertex θ, and so by the separation Theorem 4.12 there exists a nonzero vector u and a real number α such that the hyperplane

$$H' \equiv \{x : \langle x, u \rangle = \alpha\}$$

strictly separates $\{b\}$ and S in E^m. Furthermore, since $\theta \in S$ we may assume that

$$\langle b, u \rangle < \alpha < 0$$

and

$$\langle z, u \rangle > \alpha \quad \text{for all } z \in S.$$

Since the parallel hyperplane

$$H \equiv \{x : \langle x, u \rangle = 0\}$$

will support S at θ, we have

$$\langle z, u \rangle \geq 0 \quad \text{for all } z \in S,$$

so that

$$\langle Ax, u \rangle \geq 0 \quad \text{for all } x \geq \theta,$$

and

$$\langle x, uA \rangle \geq 0 \quad \text{for all } x \geq \theta.$$

We claim that this implies $uA \geq \theta$. Indeed, if the jth coordinate of uA were negative, then the inner product of uA and the jth standard basis vector e_j would also be negative, a contradiction. Therefore, we have

$$\langle b, u \rangle < 0$$

and

$$uA \geq \theta,$$

so the vector u is a solution to (5). ∎

26.9. Lemma. Exactly one of the following alternatives holds. Either the inequality

$$Ax \leq b \tag{6}$$

has a nonnegative solution, or the inequalities

$$yA \geq \theta, \quad \langle b, y \rangle < 0 \tag{7}$$

have a nonnegative solution.

PROOF. It follows easily, just as in Lemma 26.8, that both alternatives cannot hold. Thus we suppose that (6) has no nonnegative solution. This means that

$$Ax + z = b \tag{8}$$

has no nonnegative solutions. Using standard block matrix notation, (8) can be written as

$$[A \quad I]\begin{bmatrix} x \\ z \end{bmatrix} = b, \tag{9}$$

where $[A \quad I]$ is the block matrix whose left n columns are the columns of A and whose right m columns are the columns of the $m \times m$ identity matrix I, and $\begin{bmatrix} x \\ z \end{bmatrix}$ is the column vector whose first n entries are the coordinates of x and whose last m entries are the coordinates of z. Since (9) has no nonnegative solution, Lemma 26.8 implies the existence of a vector y such that

$$y[A \quad I] \geq \theta \quad \text{and} \quad \langle b, y \rangle < 0.$$

But $y[A \quad I] \geq \theta$ implies $yA \geq \theta$ and $y \geq \theta$, so we have found a nonnegative solution to (7). ∎

We are now in a position to prove the fundamental duality theorem.

PROOF OF THEOREM 26.7. (a) Let $x \in \mathcal{F}$ and $y \in \mathcal{F}^*$. That is, x and y are nonnegative solutions to the inequalities

$$Ax \leq b \tag{10}$$

$$yA \geq c \tag{11}$$

LINEAR PROGRAMMING

Part (a) will be proved if we can find solutions of (10) and (11) that also satisfy

$$\langle c, x \rangle - \langle y, b \rangle \geq 0. \tag{12}$$

Indeed, from Theorem 26.6(a) we know that $\langle c, x \rangle \leq \langle y, b \rangle$, and so if (12) holds they must be equal. But then Theorem 26.6(b) implies that x and y are optimal.

Suppose now that the system (10), (11), and (12) has no nonnegative solution. Rewriting these inequalities in block matrix notation, we have

$$\begin{bmatrix} A & 0 \\ 0 & -A^t \\ -c & b \end{bmatrix} \begin{bmatrix} x \\ y \end{bmatrix} \leq \begin{bmatrix} b \\ -c \\ 0 \end{bmatrix}, \tag{13}$$

where 0 is an appropriate sized zero matrix.

If the inequality (13) has no nonnegative solution, then Lemma 26.9 implies that there exist $z \geq \theta$ in E^m, $w \geq \theta$ in E^n and a real number $\alpha \geq 0$ such that

$$\begin{bmatrix} z & w & \alpha \end{bmatrix} \begin{bmatrix} A & 0 \\ 0 & -A^t \\ -c & b \end{bmatrix} \geq \theta$$

and

$$\begin{bmatrix} z & w & \alpha \end{bmatrix} \begin{bmatrix} b \\ -c \\ 0 \end{bmatrix} < 0.$$

This is equivalent to

$$zA \geq \alpha c \tag{14}$$

$$Aw \leq \alpha b \tag{15}$$

$$\langle z, b \rangle - \langle w, c \rangle < 0. \tag{16}$$

Now α cannot be zero, for if it were then for $x \in \mathcal{F}$ and $y \in \mathcal{F}^*$ we would get from (14), (15), (10), and (11),

$$0 \leq \langle zA, x \rangle = \langle z, Ax \rangle \leq \langle z, b \rangle$$

and

$$0 \geq \langle Aw, y \rangle = \langle w, yA \rangle \geq \langle w, c \rangle$$

which contradicts (16).

We conclude that α must be positive. Dividing equations (14) and (15) by α we see that $w/\alpha \in \mathcal{F}$ and $z/\alpha \in \mathcal{F}^*$. Hence by Theorem 26.6(a)

$$\left\langle \frac{w}{\alpha}, c \right\rangle \leq \left\langle \frac{z}{\alpha}, b \right\rangle$$

and so

$$\langle w, c \rangle \leq \langle z, b \rangle,$$

which again contradicts (16).

Thus we are forced to conclude that the original system (10), (11), and (12) has a nonnegative solution, and this solution yields optimal solutions to both P and P^*.

To prove part (b), we suppose, for example, that $\mathcal{F}$ is empty so that inequality (2) has no nonnegative solution. By Lemma 26.9, there must exist a vector $z \geq \theta$ such that

$$zA \geq \theta \quad \text{and} \quad \langle b, z \rangle < 0. \tag{17}$$

If there is a feasible solution y in $\mathcal{F}^*$, then

$$yA \geq c.$$

It follows from (17) that $y + \lambda z \in \mathcal{F}^*$ for all positive λ. Since $\langle b, z \rangle$ is negative, this implies that the inner product

$$\langle b, y + \lambda z \rangle = \langle b, y \rangle + \lambda \langle b, z \rangle$$

can be made arbitrarily small. Thus it has no minimum, and P^* has no optimal solution.

The proof of part (c) follows easily from parts (a) and (b). Indeed if, for example, P has an optimal solution $\bar{x}$, then by part (b), $\mathcal{F}^*$ is nonempty. Part (a) then implies that P^* has an optimal solution $\bar{y}$ with $f(\bar{x}) = g(\bar{y})$. ∎

26.10. Example. Let us return to Example 26.5 and look at its dual problem P^*:

$$\text{Minimize} \quad g(y_1, y_2, y_3) \equiv 30 y_1 + 20 y_2 + 52 y_3$$

$$\text{subject to} \quad y_1 + y_3 \geq 2$$

$$\phantom{\text{subject to} \quad y_1 +} y_2 + 2 y_3 \geq 3$$

and $y_1 \geq 0$, $y_2 \geq 0$, $y_3 \geq 0$.

From the Duality Theorem (26.7) we would expect P^* to have an optimal solution $\bar{y}$ such that $g(\bar{y}) = 93$. By graphing the constraint inequalities (Figure

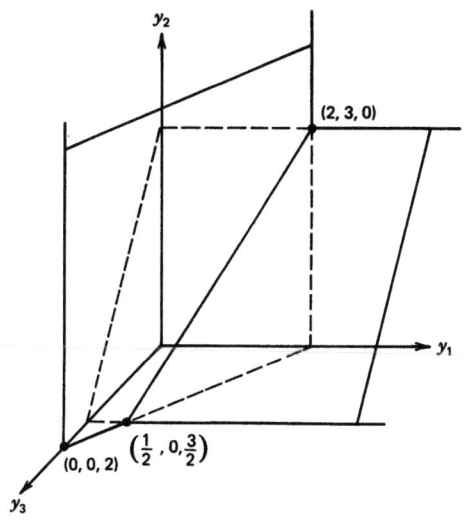

Figure 26.3.

y	$g(y)$
$(0,0,2)$	104
$(\frac{1}{2},0,\frac{3}{2})$	93 ←
$(2,3,0)$	120

26.3), we see that $\mathcal{F}^*$ has three extreme points and that $(\frac{1}{2},0,\frac{3}{2})$ gives the optimal solution.

This example also illustrates another property of duality. At the optimal solution $\bar{x} = (30, 11)$ for P, we have a strict inequality only in the second constraint:

$$\bar{x}_2 < 20.$$

Corresponding to this, the second coordinate in $\bar{y}$ is zero:

$$\bar{y}_2 = 0.$$

This is no coincidence, as the following theorem shows. The converse is also true. Noting that yA is a (row) vector, we let $(yA)_j$ denote its jth coordinate. Similarly, $(Ax)_i$ is the ith coordinate of the (column) vector Ax.

26.11. Theorem (Equilibrium Theorem). If $x \in \mathcal{F}$ and $y \in \mathcal{F}^*$, then x and y are optimal solutions to P and P^*, respectively, if and only if

$$x_j = 0 \quad \text{whenever} \quad (yA)_j > c_j \tag{18}$$

and

$$y_i = 0 \quad \text{whenever} \quad (Ax)_i < b_i. \tag{18*}$$

PROOF. To begin with, suppose that conditions (18) and (18*) hold. Then from (2*) and (18) we obtain

$$f(x) = \langle c, x \rangle = \sum_{j=1}^{n} c_j x_j$$

$$= \sum_{j=1}^{n} (yA)_j x_j = \langle yA, x \rangle.$$

Similarly, from (2) and (18*) we obtain

$$g(y) = \langle y, b \rangle = \sum_{i=1}^{m} y_i b_i$$

$$= \sum_{i=1}^{m} y_i (Ax)_i = \langle y, Ax \rangle.$$

Since $\langle yA, x \rangle = \langle y, Ax \rangle$, we have $f(x) = g(y)$, and so by Theorem 26.6(b), x and y are optimal solutions.

Conversely, if x and y are optimal solutions, then from the Duality Theorem 26.7 we know that

$$\sum_{j=1}^{n} c_j x_j = \sum_{j=1}^{n} (yA)_j x_j = \sum_{i=1}^{m} y_i (Ax)_i = \sum_{i=1}^{m} y_i b_i. \qquad (19)$$

From the equality on the left we have

$$\sum_{j=1}^{n} \left[c_j - (yA)_j \right] x_j = 0.$$

Now since $y \in \mathcal{F}^*$, none of the terms $c_j - (yA)_j$ are positive. But each $x_j \geq 0$. Thus we must have

$$\left[c_j - (yA)_j \right] x_j = 0$$

for each j. Condition (18) follows immediately from this, and condition (18*) follows in a similar manner from the equality on the right in (19). ∎

We have already observed that there is a similarity between linear programming and game theory. In fact, the overall strategy for proving the main theorems in each case is parallel. (Compare Lemmas 25.8 and 25.9 with Lemmas 26.8 and 26.9.) We conclude this section by showing that the Fundamental Theorem of Matrix Games (Theorem 25.7) can actually be

LINEAR PROGRAMMING

derived from the theory of linear programming. (See Section 25 for the necessary definitions and notation.)

26.12. Alternate Proof of Theorem 25.7. We begin by supposing that each entry in the payoff matrix A is positive. Let $u \in \mathbf{E}^m$ and $v \in \mathbf{E}^n$ be the vectors whose coordinates are all equal to one. Consider the following linear program P and its dual P^*. (Note that the roles of x and y are interchanged. Thus $x \in \mathbf{E}^m$ and $y \in \mathbf{E}^n$.)

P: Maximize $\langle y, v \rangle$ P^*: Minimize $\langle x, u \rangle$

subject to $Ay \leq u$ subject to $xA \geq v$

and $y \geq \theta$ and $x \geq \theta$.

The primal problem P is feasible since $y = \theta$ satisfies the constraints. The dual problem P^* is feasible since $A > 0$. Thus the Duality Theorem (26.7) implies the existence of optimal solutions x_0 and y_0 such that

$$\langle x_0, u \rangle = \langle y_0, v \rangle \equiv \lambda.$$

It is clear that $\lambda > 0$, and we claim that

$$\bar{x} \equiv \frac{x_0}{\lambda} \quad \text{and} \quad \bar{y} \equiv \frac{y_0}{\lambda} \tag{20}$$

are optimal mixed strategies for the matrix game A. Furthermore, the value of the game, μ, is equal to $1/\lambda$.

The fact that $\bar{x}$ and $\bar{y}$ are mixed strategies follows from (20) and the definitions of u and v. Now if y is any mixed strategy, then $\langle v, y \rangle = 1$. Since $x_0 A \geq v$, we have

$$\langle \bar{x}, Ay \rangle = \langle \bar{x}A, y \rangle$$

$$= \frac{1}{\lambda} \langle x_0 A, y \rangle$$

$$\geq \frac{1}{\lambda} \langle v, y \rangle = \frac{1}{\lambda}.$$

Hence $E(\bar{x}, y) \geq 1/\lambda$ for all $y \in Y$, and it follows that $v_{\mathrm{I}} \geq 1/\lambda$. Similarly, we find $E(x, \bar{y}) \leq 1/\lambda$ for all $x \in X$ and so $v_{\mathrm{II}} \leq 1/\lambda$. Hence by Theorem 25.4, $v_{\mathrm{I}} = 1/\lambda = v_{\mathrm{II}}$ so that $\mu \equiv 1/\lambda$ is the value of the game. Since $E(\bar{x}, y) \geq \mu$ for all $y \in Y$ and $E(x, \bar{y}) \leq \mu$ for all $x \in X$, $\bar{x}$ and $\bar{y}$ are the optimal strategies.

Having now completed the proof for positive matrices, suppose that $B \equiv (b_{ij})$ is an arbitrary $m \times n$ matrix. Choose a real number β such that

$$\beta < \min_{i,j} b_{ij},$$

and let $A \equiv (a_{ij})$ be the matrix given by

$$a_{ij} = b_{ij} - \beta.$$

It follows that $A > 0$, and so by applying the preceding argument, we obtain optimal strategies $\bar{x}$ and $\bar{y}$ such that

$$\langle \bar{x}, Ay \rangle \geqslant \mu \geqslant \langle xA, \bar{y} \rangle \qquad \text{for all } x \in X, y \in Y.$$

Since the sum of the coordinates of any x or y is equal to one, it follows readily that $\langle \bar{x}, Ay \rangle = \langle \bar{x}, By \rangle - \beta$ and $\langle xA, \bar{y} \rangle = \langle xB, \bar{y} \rangle - \beta$. Hence we have

$$\langle \bar{x}, By \rangle \geqslant \mu + \beta \geqslant \langle xB, \bar{y} \rangle \qquad \text{for all } x \in X, y \in Y.$$

Thus $(\bar{x}, \bar{y})$ is a solution of the matrix game B, and the game has a value of $\mu + \beta$. ∎

EXERCISES

26.1. Solve the following linear programming problems.
    (a) Maximize $80x_1 + 70x_2$
        subject to $\quad 2x_1 + x_2 \leqslant 32$
        $\qquad\qquad\quad x_1 + x_2 \leqslant 18$
        $\qquad\qquad\quad x_1 + 3x_2 \leqslant 24$
    and $x_1 \geqslant 0, x_2 \geqslant 0$.
    (b) Minimize $5y_1 + 4y_2$
        subject to $\quad 2y_1 + 5y_2 \geqslant 10$
        $\qquad\qquad\quad 3y_1 + y_2 \geqslant 6$
        $\qquad\qquad\quad y_1 + 7y_2 \geqslant 7$
    and $y_1 \geqslant 0, y_2 \geqslant 0$.
    (c) Maximize $3x_1 + 4x_2$
        subject to $\quad -2x_1 + x_2 \leqslant -4$
        $\qquad\qquad\quad x_1 - 2x_2 \leqslant -4$
    and $x_1 \geqslant 0, x_2 \geqslant 0$.
    (d) Maximize $2x_1 + 5x_2$
        subject to $\quad 2x_1 - 2x_2 \leqslant 6$
        $\qquad\qquad\quad -x_1 + 2x_2 \leqslant -4$
    and $x_1 \geqslant 0, x_2 \geqslant 0$.
    (e) Minimize $3y_1 + 2y_2$
        subject to $\quad 2y_1 - y_2 \geqslant 4$
        $\qquad\qquad\quad -y_1 + 2y_2 \geqslant 4$
    and $y_1 \geqslant 0, y_2 \geqslant 0$.

26.2. Consider the primal problem P:
        Maximize $9x_1 + 10x_2$
        subject to $\quad 5x_1 + 4x_2 \leqslant 80$
        $\qquad\qquad\quad 5x_1 + 8x_2 \leqslant 120$
    and $x_1 \geqslant 0, x_2 \geqslant 0$.

THE SIMPLEX METHOD

      (a) Solve P.
      (b) State the dual minimizing problem P^*.
      (c) Solve P^*.

26.3. State and solve the duals to Exercises 26.1(c)–(e).

26.4. Let A be an $m \times n$ matrix and let $S \equiv \{z: z = Ax, x \geqslant \theta\}$. Prove that S is a closed convex cone with vertex θ. (See Exercise 5.15 for the definition.)

26.5. If the primal linear programming problem P is unbounded, prove that the dual problem P^* is infeasible.

26.6. Show by an example that it is possible for both the primal linear programming problem P and its dual P^* to be infeasible.

26.7. Consider the matrix game $A \equiv \begin{bmatrix} 1 & 3 \\ 5 & 2 \end{bmatrix}$.
      (a) Set up the primal and dual linear programming problems necessary to find the optimal strategies for A.
      (b) Solve the programs in (a) and determine the optimal strategies for A.
      (c) Check your answer to (b) by solving A directly as a matrix game.

26.8. A toy manufacturing company makes two kinds of toys: invertible widgets and collapsible whammies. The manufacturing process for each widget requires 12 cubic feet of natural gas, 10 kilowatt hours of electricity, and 10 gallons of fuel oil. Each whammy requires 12 cubic feet of natural gas, 20 kilowatt hours of electricity, and 4 gallons of fuel oil. Due to a fuel shortage the company is allotted only 600 cubic feet of natural gas, 800 kilowatt hours of electricity, and 400 gallons of fuel oil per week. The profit on each widget is \$2.00 and on each whammy is \$2.70.
      (a) How many of each toy should they make each week to maximize their profit?
      (b) Suppose that the demand for collapsible whammies suddenly declined so that the toy company had to lower the price (and hence lower the profit) in order to sell them. If other factors remained the same, how small would their profit on each whammy have to be before they would switch production entirely to widgets?

SECTION 27. THE SIMPLEX METHOD

In Section 26 we described a graphical technique for solving linear programming problems in two variables. In actual applications in business and industry, it is not uncommon to encounter dozens or even hundreds of variables. Thus it is important to have a more general procedure available. In this section we describe the details of the simplex method (or simplex algorithm). It is not restricted in the number of variables it can handle, and because of its highly structured format, it is readily adapted to use on a computer. By virtue of the relationship between linear programming and game theory (see Proof 26.12), it can also be used to solve arbitrarily large two-person matrix games.

We begin our discussion of the simplex method by restricting our attention to the canonical linear programming problem in which each of the coordinates in the m-tuple $b \equiv (b_1, \ldots, b_m)$ is positive. Thus we have:

$$\text{Maximize} \quad f(x) \equiv \langle c, x \rangle$$

subject to the constraints

$$Ax \leq b$$

and $x \geq \theta$,

where $x \equiv (x_1, \ldots, x_n)$, $A \equiv (a_{ij})$, $C \equiv (c_1, \ldots, c_n)$, and $b \equiv (b_1, \ldots, b_m) > \theta$.

In order to use the simplex method we must change each of the constraint inequalities in $Ax \leq b$ into an equality. We do this by adding new variables.

27.1. Definition. A nonnegative variable that is added to the smaller side of an inequality to convert it to an equality is called a **slack variable**.

For example, the inequality

$$2x_1 + 3x_2 \leq 50$$

would be changed to the equality

$$2x_1 + 3x_2 + x_3 = 50,$$

by adding the slack variable x_3, where $x_3 \equiv 50 - (2x_1 + 3x_2) \geq 0$.

After introducing the slack variables in each equation, we obtain a linear system with m equations and $m + n$ variables. We say that a solution to this system is a **basic solution** if no more than m of the variables are nonzero. Corresponding to our earlier terminology, a solution to the system is called **feasible** if each variable is nonnegative. Thus in a **basic feasible solution**, each variable must be nonnegative and at most m of them can be positive. Geometrically, these basic feasible solutions correspond to the extreme points of the feasible region.

27.2. Example. Consider the following constraint inequalities:

$$2x_1 + 3x_2 + 4x_3 \leq 60$$

$$3x_1 + x_2 + 5x_3 \leq 45$$

$$x_1 + 2x_2 + x_3 \leq 50.$$

THE SIMPLEX METHOD

By adding slack variables we obtain the equations,

$$2x_1 + 3x_2 + 4x_3 + x_4 = 60$$
$$3x_1 + x_2 + 5x_3 + x_5 = 45$$
$$x_1 + 2x_2 + x_3 + x_6 = 50.$$

There is a simple basic feasible solution associated with this system of equations. Since the variables x_4, x_5, and x_6 each occur in only one equation (and each have a coefficient of 1), we can let the other variables all be equal to zero and easily solve the system. Thus we say that the basic feasible solution associated with the system is

$$x_1 = x_2 = x_3 = 0, \quad x_4 = 60, \quad x_5 = 45, \quad \text{and} \quad x_6 = 50.$$

It is customary to refer to the nonzero variables x_4, x_5, and x_6 as **basic variables** that are "in" this solution. The variables x_1, x_2, and x_3 are said to be "out" of the solution. In a linear programming problem this particular solution would probably not be optimal since only the slack variables are nonzero.

Suppose that we wish to introduce another variable, say x_2, "into" the solution. (For the moment we shall avoid the question of why we chose x_2.) This may be accomplished by using the Gauss-Jordan procedure from linear algebra.

Specifically, we shall pivot on the x_2 entry in the third equation. To do this we divide the third equation by 2 (the coefficient a_{32} of x_2) to obtain a new third equation:

$$\tfrac{1}{2}x_1 + x_2 + \tfrac{1}{2}x_3 + \tfrac{1}{2}x_6 = 25.$$

We multiply this new third equation by -3 (the negative of the coefficient a_{12} of x_2 in the first equation) and add it to the first equation to obtain a new first equation:

$$\tfrac{1}{2}x_1 + \tfrac{5}{2}x_3 + x_4 - \tfrac{3}{2}x_6 = -15.$$

Similarly, we multiply the new third equation by -1 (the negative of a_{22}) and add it to the second equation to obtain a new second equation. Our three equations now read:

$$\tfrac{1}{2}x_1 + \tfrac{5}{2}x_3 + x_4 - \tfrac{3}{2}x_6 = -15$$
$$\tfrac{5}{2}x_1 + \tfrac{9}{2}x_3 + x_5 - \tfrac{1}{2}x_6 = 20$$
$$\tfrac{1}{2}x_1 + x_2 + \tfrac{1}{2}x_3 + \tfrac{1}{2}x_6 = 25.$$

This new system gives us the basic solution:

$$x_1 = x_3 = x_6 = 0, \quad x_2 = 25, \quad x_4 = -15, \quad x_5 = 20.$$

We have brought the variable x_2 into the solution and the variable x_6 has gone out.

Unfortunately, this basic solution is not feasible since $x_4 < 0$. What has gone wrong? When we brought the variable x_2 into the solution, we pivoted on the wrong equation.

In general, suppose we have the system

$$a_{11}x_1 + \cdots + a_{1k}x_k + \cdots + a_{1n}x_n = b_1$$
$$\vdots$$
$$a_{i1}x_1 + \cdots + a_{ik}x_k + \cdots + a_{in}x_n = b_i$$
$$\vdots$$
$$a_{m1}x_1 + \cdots + a_{mk}x_k + \cdots + a_{mn}x_n = b_m,$$

and we wish to bring the variable x_k into the solution. If we pivot on the pth equation, then the resulting system will yield a basic feasible solution if the following two conditions are satisfied:

1. The coefficient a_{pk} of x_k must be positive. (We want to divide the pth equation by a_{pk} and have the new b_p term be positive.)

2. We choose p so that $b_p/a_{pk} < b_i/a_{ik}$ for all i for which $a_{ik} > 0$. (This will guarantee that when we combine the pth equation with each of the other equations in order to cancel out the x_k term, the resulting b_i term will be positive.)

Returning to Example 27.2, we see that

$$\frac{b_1}{a_{12}} = \frac{60}{3} = 20, \quad \frac{b_2}{a_{22}} = 45, \quad \text{and} \quad \frac{b_3}{a_{32}} = \frac{50}{2} = 25.$$

Thus in order to bring x_2 into the solution we should pivot on the x_2 term in the *first* equation to obtain the system:

$$\tfrac{2}{3}x_1 + x_2 + \tfrac{4}{3}x_3 + \tfrac{1}{3}x_4 = 20$$
$$\tfrac{7}{3}x_1 \quad\quad + \tfrac{11}{3}x_3 - \tfrac{1}{3}x_4 + x_5 = 25$$
$$-\tfrac{1}{3}x_1 \quad\quad - \tfrac{5}{3}x_3 - \tfrac{2}{3}x_4 \quad\quad + x_6 = 10,$$

which yields the basic feasible solution

$$x_1 = x_3 = x_4 = 0, \quad x_2 = 20, \quad x_5 = 25, \quad x_6 = 10.$$

THE SIMPLEX METHOD

Our work will be simplified considerably if at this point we introduce the matrix form for a system of linear equations. Our first system of equations in Example 27.2 can be abbreviated by the augmented matrix

$$\begin{array}{cccccc} x_1 & x_2 & x_3 & x_4 & x_5 & x_6 \end{array}$$
$$\begin{bmatrix} 2 & ③ & 4 & 1 & 0 & 0 & | & 60 \\ 3 & 1 & 5 & 0 & 1 & 0 & | & 45 \\ 1 & 2 & 1 & 0 & 0 & 1 & | & 50 \end{bmatrix},$$

where we have labeled the columns corresponding to the variables $x_1, \ldots, x_6$. The basic feasible solution associated with this matrix is

$$x_1 = x_2 = x_3 = 0, \quad x_4 = 60, \quad x_5 = 45, \quad x_6 = 50.$$

We have circled the entry 3 in the x_2 column to indicate that this will be used as a pivot to bring x_2 into the solution. After applying the Gauss-Jordan procedure we obtain the new matrix

$$\begin{array}{cccccc} x_1 & x_2 & x_3 & x_4 & x_5 & x_6 \end{array}$$
$$\begin{bmatrix} \frac{2}{3} & 1 & \frac{4}{3} & \frac{1}{3} & 0 & 0 & | & 20 \\ \frac{7}{3} & 0 & \frac{11}{3} & -\frac{1}{3} & 1 & 0 & | & 25 \\ -\frac{1}{3} & 0 & -\frac{5}{3} & -\frac{2}{3} & 0 & 1 & | & 10 \end{bmatrix},$$

and our new basic feasible solution is again found to be

$$x_1 = x_3 = x_4 = 0, \quad x_2 = 20, \quad x_5 = 25, \quad x_6 = 10.$$

Having now described the technique of bringing a particular variable into the solution, let us consider the question of which variable (if any) should be brought into the solution at a particular point. To do this we introduce an objective function to be maximized in our previous example.

27.3. Example

$$\text{Maximize} \quad 25x_1 + 30x_2 + 18x_3$$

$$\text{subject to} \quad 2x_1 + 3x_2 + 4x_3 + x_4 \qquad\qquad = 60$$

$$3x_1 + x_2 + 5x_3 \qquad + x_5 \qquad = 45$$

$$x_1 + 2x_2 + x_3 \qquad\qquad + x_6 = 50$$

$$\text{and } x_j \geq 0 \quad \text{for } j = 1, \ldots, 6.$$

SOLUTION. We begin by changing our objective function $25x_1 + 30x_2 + 18x_3$ into an equation by introducing a new variable M defined by $M \equiv 25x_1 + 30x_2 + 18x_3$. Then M is the variable we want to maximize and it satisfies the equation

$$-25x_1 - 30x_2 - 18x_3 + M = 0.$$

Thus our problem can be restated as follows: Among all the solutions of the system of equations

$$\begin{aligned} 2x_1 + 3x_2 + 4x_3 + x_4 &= 60 \\ 3x_1 + x_2 + 5x_3 \phantom{{}+x_4} + x_5 &= 45 \\ x_1 + 2x_2 + x_3 \phantom{{}+x_4+x_5} + x_6 &= 50 \\ -25x_1 - 30x_2 - 18x_3 \phantom{{}+x_4+x_5+x_6} + M &= 0 \end{aligned}$$

find a solution for which each $x_j \geq 0$ ($i = 1, \ldots, 6$) and for which M is as large as possible.

When we write this new system in matrix form, we obtain what is called the **initial simplex tableau**:

$$\begin{array}{ccccccc} x_1 & x_2 & x_3 & x_4 & x_5 & x_6 & M \end{array}$$
$$\left[\begin{array}{ccccccc|c} 2 & ③ & 4 & 1 & 0 & 0 & 0 & 60 \\ 3 & 1 & 5 & 0 & 1 & 0 & 0 & 45 \\ 1 & 2 & 1 & 0 & 0 & 1 & 0 & 50 \\ \hline -25 & -30 & -18 & 0 & 0 & 0 & 1 & 0 \end{array}\right]$$

Notice that we have drawn a horizontal line above the bottom row. This was done to separate off the equation corresponding to the objective function. This last row will play a special role in what follows. (We never pivot on an entry in the bottom row, but we use it to decide which variable to bring into the solution.)

Associated with the tableau above we have the basic feasible solution

$$x_1 = x_2 = x_3 = 0, \quad x_4 = 60, \quad x_5 = 45, \quad x_6 = 50, \quad M = 0.$$

This solution is not optimal, however, since from the bottom row we obtain

$$M = 25x_1 + 30x_2 + 18x_3.$$

Thus we can increase the value of M by increasing any of the variables x_1, x_2, or x_3. Since the coefficient of x_2 is the largest of the three coefficients, we can

THE SIMPLEX METHOD

increase M most rapidly by increasing x_2. That is to say, it is to our best advantage to bring x_2 into the solution.

We now bring x_2 into the solution using the Gauss-Jordan procedure outlined earlier. By comparing the ratios b_i/a_{i2} for each row except the last, we decide to pivot on the entry 3 that is circled in the first row. The result is

$$\begin{array}{ccccccc|c} x_1 & x_2 & x_3 & x_4 & x_5 & x_6 & M & \\ \frac{2}{3} & 1 & \frac{4}{3} & \frac{1}{3} & 0 & 0 & 0 & 20 \\ \boxed{\frac{7}{3}} & 0 & \frac{11}{3} & -\frac{1}{3} & 1 & 0 & 0 & 25 \\ -\frac{1}{3} & 0 & -\frac{5}{3} & -\frac{1}{3} & 0 & 1 & 0 & 10 \\ \hline -5 & 0 & 22 & 10 & 0 & 0 & 1 & 600 \end{array}$$

The corresponding basic feasible solution is

$$x_1 = x_3 = x_4 = 0, \quad x_2 = 20, \quad x_5 = 25, \quad x_6 = 10, \quad M = 600.$$

Thus we have increased M from 0 to 600. Can M be increased further? To answer this we look at the bottom row of the tableau and solve the equation for M:

$$M = 600 + 5x_1 - 22x_3 - 10x_4.$$

Since each of the variables x_j is nonnegative, we can increase the value of M by increasing x_1. Since the coefficients of x_3 and x_4 are both negative at this point, it is to our advantage to leave them equal to zero. Having decided to bring x_1 into the solution, we compare the ratios

$$\frac{20}{\frac{2}{3}} = 30 \quad \text{and} \quad \frac{25}{\frac{7}{3}} = \frac{75}{7}$$

and decide to pivot on the $\frac{7}{3}$. The resulting tableau is

$$\begin{array}{ccccccc|c} x_1 & x_2 & x_3 & x_4 & x_5 & x_6 & M & \\ 0 & 1 & \frac{2}{7} & \frac{3}{7} & -\frac{2}{7} & 0 & 0 & \frac{90}{7} \\ 1 & 0 & \frac{11}{7} & -\frac{1}{7} & \frac{3}{7} & 0 & 0 & \frac{75}{7} \\ 0 & 0 & \frac{8}{7} & -\frac{8}{21} & \frac{3}{7} & 1 & 0 & \frac{95}{7} \\ \hline 0 & 0 & \frac{209}{7} & \frac{65}{7} & \frac{15}{7} & 0 & 1 & \frac{4575}{7} \end{array}$$

The corresponding basic feasible solution is

$$x_3 = x_4 = x_5 = 0, \quad x_1 = \frac{75}{7}, \quad x_2 = \frac{90}{7}, \quad x_6 = \frac{95}{7}, \quad M = \frac{4575}{7}.$$

From the bottom row we obtain

$$M = \tfrac{4575}{7} - \tfrac{209}{7}x_3 - \tfrac{65}{7}x_4 - \tfrac{15}{7}x_5.$$

Since each of the variables x_j is nonnegative, we see that M can be no larger than $\tfrac{4575}{7}$. But our present solution has M equal to $\tfrac{4575}{7}$, so we conclude that it must be optimal.

Notice that a *negative* entry in the bottom row of any x_j column will become a *positive* coefficient when that equation is solved for M. Thus we know we have reached the optimal solution when each of these entries in the tableau is nonnegative.

We now summarize the simplex method for solving the canonical maximizing problem in which each coordinate of the vector b is positive.

1. Change the inequality constraints into equalities by adding slack variables. Let M be a variable equal to the objective function and write the corresponding equation on the bottom.
2. Set up the initial simplex tableau. The slack variables (and M) provide the initial basic feasible solution.
3. Check the bottom row of the tableau for optimality. If all the entries to the left of the vertical line are nonnegative, then the solution is optimal. If some are negative, then choose the variable x_k for which the entry in the bottom row is as negative as possible.
4. Bring the variable x_k into the solution. Do this by pivoting on the positive entry a_{pk} for which the ratio b_i/a_{ik} is the smallest. By so doing we will obtain a new basic feasible solution and the value of M will be increased.
5. Repeat the process, beginning at step 3, until all the entries in the bottom row are nonnegative.

Two things can go wrong in this process. At step 4 we might have a negative entry in the bottom row of the x_k column, but there might be no positive a_{ik} entry above it. In this case it will not be possible to find a pivot to bring x_k into the solution. This corresponds to the case where the objective function is unbounded, and no optimal solution will exist.

The second potential problem also occurs at step 4. The smallest ratio b_p/a_{pk} may occur in more than one row. When this happens the following tableau will have at least one basic variable equal to zero, and in subsequent tableaus the value of M may not necessarily be increased. (It may remain constant.) Theoretically it is possible for an infinite sequence of pivots to occur and fail to lead to an optimal solution. Such a phenomenon is called **cycling**. Fortunately, cycling does not seem to occur in practical applications, and we will not concern ourselves with it further. The interested reader is referred to the article on degeneracy by Charnes (1952). In practice, when the smallest ratio b_p/a_{pk} occurs in more than one row we may arbitrarily choose either of these rows to pivot.

THE SIMPLEX METHOD

27.4. Example. A health food store sells two different mixtures of nuts. A box of the first mixture contains 1 pound of cashews and 1 pound of peanuts. A box of the second mixture contains 1 pound of filberts and 2 pounds of peanuts. The store has available 30 pounds of cashews, 20 pounds of filberts, and 52 pounds of peanuts. Suppose the profit on each box of the first mixture is $2 and on each box of the second mixture is $3. If the store can sell all of the boxes it mixes, how many boxes of each mixture should be made in order to maximize the profit?

SOLUTION. If we let x_1 be the number of boxes of the first mixture and x_2 be the number of boxes of the second mixture, then the problem can be expressed mathematically as

$$\text{Maximize} \quad 2x_1 + 3x_2$$

$$\text{subject to} \quad x_1 \leq 30$$

$$x_2 \leq 20$$

$$x_1 + 2x_2 \leq 52$$

and $x_1 \geq 0$, $x_2 \geq 0$.

This turns out to be the same problem we solved graphically in Example 26.5. Let us now solve it by the simplex method and relate the basic feasible solution from each tableau to the corresponding extreme point of the feasible region. (See Figure 26.1.)

Adding slack variables and rewriting the objective function, we obtain the equivalent problem of finding a nonnegative solution to the system

$$x_1 \quad\quad + x_3 \quad\quad\quad\quad\quad\quad = 30$$

$$x_2 \quad\quad + x_4 \quad\quad\quad\quad = 20$$

$$x_1 + 2x_2 \quad\quad\quad\quad + x_5 \quad\quad = 52$$

$$-2x_1 - 3x_2 \quad\quad\quad\quad\quad\quad + M = 0$$

for which M is a maximum. Our initial simplex tableau is

$$\begin{array}{cccccc|c}
x_1 & x_2 & x_3 & x_4 & x_5 & M & \\
\hline
1 & 0 & 1 & 0 & 0 & 0 & 30 \\
0 & ① & 0 & 1 & 0 & 0 & 20 \\
1 & 2 & 0 & 0 & 1 & 0 & 52 \\
\hline
-2 & -3 & 0 & 0 & 0 & 1 & 0
\end{array}.$$

The basic feasible solution is

$$x_1 = x_2 = 0, \quad x_3 = 30, \quad x_4 = 20, \quad x_5 = 52, \quad M = 0,$$

which corresponds to the extreme point $(x_1, x_2) = (0, 0)$ of the feasible region. Pivoting on the 1 in the x_2 column, we obtain

$$\begin{array}{cccccc} x_1 & x_2 & x_3 & x_4 & x_5 & M \end{array}$$
$$\left[\begin{array}{cccccc|c} 1 & 0 & 1 & 0 & 0 & 0 & 30 \\ 0 & ① & 0 & 1 & 0 & 0 & 20 \\ 1 & 0 & 0 & -2 & 1 & 0 & 12 \\ \hline -2 & 0 & 0 & 3 & 0 & 1 & 60 \end{array}\right]$$

The basic feasible solution is now

$$x_1 = x_4 = 0, \quad x_2 = 20, \quad x_3 = 30, \quad x_5 = 12, \quad M = 60.$$

Thus we have moved up to the extreme point $(x_1, x_2) = (0, 20)$. Bringing x_1 into the solution gives us

$$\begin{array}{cccccc} x_1 & x_2 & x_3 & x_4 & x_5 & M \end{array}$$
$$\left[\begin{array}{cccccc|c} 0 & 0 & 1 & ② & -1 & 0 & 18 \\ 0 & 1 & 0 & 1 & 0 & 0 & 20 \\ 1 & 0 & 0 & -2 & 1 & 0 & 12 \\ \hline 0 & 0 & 0 & -1 & 2 & 1 & 84 \end{array}\right]$$

This time $x_1 = 12$ and $x_2 = 20$, so we have moved across to the extreme point $(12, 20)$. The objective function has increased to 84. Our final tableau reads

$$\begin{array}{cccccc} x_1 & x_2 & x_3 & x_4 & x_5 & M \end{array}$$
$$\left[\begin{array}{cccccc|c} 0 & 0 & \frac{1}{2} & 1 & -\frac{1}{2} & 0 & 9 \\ 0 & 1 & -\frac{1}{2} & 0 & \frac{1}{2} & 0 & 11 \\ 1 & 0 & 1 & 0 & 0 & 0 & 30 \\ \hline 0 & 0 & \frac{1}{2} & 0 & \frac{3}{2} & 1 & 93 \end{array}\right]$$

Since all the entries in the bottom row are nonnegative, we have reached the optimal solution. We find $x_1 = 30$ and $x_2 = 11$, corresponding to the extreme point $(30, 11)$. Thus our maximum profit of \$93 will be attained by making 30 boxes of the first mixture and 11 boxes of the second.

Our final tableau in Example 27.4 also exhibits another remarkable property of the simplex method: Not only have we solved the original maximizing

THE SIMPLEX METHOD

problem, but also we have found the optimal solution to the dual minimizing problem! The variables x_3, x_4, and x_5 are the slack variables for the first, second, and third equations, respectively. The bottom entries in each of these columns give us the optimal solution $\bar{y} \equiv (\frac{1}{2}, 0, \frac{3}{2})$ to the dual problem. (See Example 26.10.) We see that the second coordinate of $\bar{y}$ is zero since x_4 (the slack variable in the second equation) is a basic variable in our final tableau. This corresponds to the second inequality in our original problem being a strict inequality at the optimal solution. We observed this same phenomenon earlier when we proved the equilibrium theorem (Theorem 26.11).

We now know how to solve the canonical maximizing problem when each of the coordinates of the vector b is positive. But what if some of the coordinates of b are zero or negative? And what about a minimization problem?

If some of the coordinates of b are zero, then theoretically cycling can occur and the simplex method will not lead us to an optimal solution. As mentioned earlier, this does not seem to happen in practical applications. Thus the presence of zero entries in the right-hand column causes no real problems in the operation of the simplex method as described.

Unfortunately, if one of the coordinates of b is negative, then the difficulty cannot be dismissed so easily. It is necessary for the b_i terms to be nonnegative in order for the slack variables to give us an initial basic feasible solution. One way to change a negative b_i term into a positive would be to multiply the inequality by -1. But this would change the direction of the inequality. For example,

$$x_1 - 2x_2 + x_3 \leq -5$$

would become

$$-x_1 + 2x_2 - x_3 \geq 5.$$

Thus the difficulty caused by a negative b_i term is the same as the difficulty caused by a reversed inequality—as would be expected in a minimization problem. We illustrate the technique used to overcome these difficulties in the following example.

27.5. Example

$$\text{Minimize} \quad x_1 + 2x_2$$

$$\text{subject to} \quad x_1 + x_2 \geq 12$$

$$x_1 - x_2 \leq 2$$

$$\text{and } x_1 \geq 0, x_2 \geq 0.$$

SOLUTION. We begin by changing the first two constraints into the "$\leq$" form, and then, instead of minimizing $x_1 + 2x_2$, we maximize its negative,

$-x_1 - 2x_2$. Thus the original problem is equivalent to the following:

$$\text{Maximize} \quad -x_1 - 2x_2$$

$$\text{subject to} \quad -x_1 - x_2 \leq -12$$

$$x_1 - x_2 \leq 2$$

and $x_1 \geq 0, x_2 \geq 0$.

Letting $M = -x_1 - 2x_2$ and adding slack variables we obtain the linear system

$$-x_1 - x_2 + x_3 \qquad\qquad = -12$$

$$x_1 - x_2 \qquad + x_4 \qquad = 2$$

$$x_1 + 2x_2 \qquad\qquad + M = 0.$$

We wish to find a nonnegative solution to this system for which M is a maximum, and to this end we construct the initial simplex tableau:

$$\begin{array}{ccccc} x_1 & x_2 & x_3 & x_4 & M \end{array}$$

$$\left[\begin{array}{ccccc|c} -1 & \ominus 1 & 1 & 0 & 0 & -12 \\ 1 & -1 & 0 & 1 & 0 & 2 \\ \hline 1 & 2 & 0 & 0 & 1 & 0 \end{array}\right].$$

The corresponding basic solution is

$$x_1 = x_2 = 0, \quad x_3 = -12, \quad x_4 = 2, \quad M = 0.$$

However, since x_3 is negative, this basic solution is not feasible. Thus before we can begin the simplex method, we have to change each of the terms in the last column above the horizontal line into a nonnegative number. (It is all right for the bottom term in the last column to be negative since it corresponds to the current value of the objective function. This possibility doesn't occur in a maximizing problem.) Fortunately, this change can be accomplished by pivoting in an appropriate way.

In order to replace a negative b_i entry by a positive number, we find another entry in the same row which is negative. (If all the other entries in the row are nonnegative, then the problem has no feasible solution. See Exercise 27.2.) This negative entry is in the column corresponding to the variable we should now bring into the solution. In our example, the entries in the first two columns are negative, so we can bring either x_1 or x_2 into the solution.

Suppose we decide to bring x_2 into the solution. In choosing the proper pivot we now select the entry a_{i2} for which the ratio b_i/a_{i2} is the smallest nonnegative number. (Note that the ratio can be positive by both b_i and a_{i2} being negative. Also note that we do not compute the ratio for the bottom row since it corresponds to the objective function. We never want to pivot on an entry in the bottom row.)

In this case only the ratio $-12/-1$ is nonnegative, and so we select the -1 in the first row as the pivot. The resulting tableau is

$$\begin{array}{c c c c c} x_1 & x_2 & x_3 & x_4 & M \end{array}$$
$$\begin{bmatrix} 1 & 1 & -1 & 0 & 0 & | & 12 \\ ② & 0 & -1 & 1 & 0 & | & 14 \\ \hline -1 & 0 & 2 & 0 & 1 & | & -24 \end{bmatrix}.$$

We find that each term in the last column (except the bottom one) is positive and we can now proceed with the usual simplex method. (Sometimes it may be necessary to pivot more than once in order to make each of these terms nonnegative. See Exercise 27.3.) Our next tableau turns out to be optimal:

$$\begin{array}{c c c c c} x_1 & x_2 & x_3 & x_4 & M \end{array}$$
$$\begin{bmatrix} 0 & 1 & -\tfrac{1}{2} & -\tfrac{1}{2} & 0 & | & 5 \\ 1 & 0 & -\tfrac{1}{2} & \tfrac{1}{2} & 0 & | & 7 \\ \hline 0 & 0 & \tfrac{3}{2} & \tfrac{1}{2} & 1 & | & -17 \end{bmatrix}.$$

Thus the maximum feasible value of $-x_1 - 2x_2$ is -17 and this occurs when $x_1 = 7$ and $x_2 = 5$. We conclude that the *minimum* value of $x_1 + 2x_2$ is 17.

Our approach to solving the minimization problem in Example 27.5 was to change it into the problem of maximizing the negative of the original objective function. Since the simplex method solves both the primal and the dual problems, an alternative approach is to solve the dual maximizing problem. This technique is particularly advantageous when the original constraint inequalities are all of the "$\geqslant$" form or when there are more of these inequalities than there are variables. (See Exercise 27.5.)

EXERCISES

27.1. Solve the following linear programming problems by using the simplex method.
    (a) Maximize $3x_1 + 4x_2$
        subject to $\quad x_1 + 2x_2 \leqslant 25$
                    $2x_1 + 3x_2 \leqslant 30$
                    $x_1 + x_2 \leqslant 14$
    and $x_1 \geqslant 0, x_2 \geqslant 0$.
    (b) Maximize $2x_1 + 5x_2 + 4x_3$
        subject to $\quad x_1 + 2x_2 + \leqslant 30$
                    $x_1 + + 2x_3 \leqslant 8$
                       $x_2 + x_3 \leqslant 12$
    and $x_j \geqslant 0$ for $j = 1, 2, 3$.

(c) Maximize $21x_1 + 25x_2 + 18x_3 + 28x_4 + 12x_5$
subject to
$2x_1 + x_2 + 3x_3 + x_4 + 2x_5 \leq 25$
$x_1 + 5x_2 + 2x_3 + x_4 + 3x_5 \leq 18$
$x_1 + 3x_2 + x_3 + 2x_4 + 2x_5 \leq 20$
and $x_j \geq 0$ for $j = 1, \ldots, 5$.

(d) Minimize $x_1 + x_2 + x_3$
subject to
$x_1 - 2x_2 \geq -9$
$2x_2 + x_3 \geq 15$
$2x_1 - x_2 + x_3 \leq 25$
and $x_j \geq 0$ for $j = 1, 2, 3$.

27.2. Suppose a simplex tableau has a negative entry for some b_i (other than the last one) and all the other entries in that row are nonnegative. Prove that the corresponding linear programming problem is infeasible.

27.3. Solve Example 27.5 by pivoting in the first tableau on the -1 entry in the first column instead of the second.

27.4. The bottom row of the final tableau for a linear programming problem will have zero entries in the columns corresponding to the basic variables. There may also be zeros in some of the other columns. When this happens the optimal solution will not be unique since these other variables could be brought into the solution without changing the objective function. Find *all* of the optimal solutions to the following problem:

Maximize $4x_1 + 5x_2 - x_3$

subject to $x_1 + 2x_2 - x_3 \leq 16$

$x_1 + x_2 \leq 12$

$2x_1 + 2x_2 + x_3 \leq 36$

and $x_j \geq 0$ for $j = 1, 2, 3$.

27.5. Suppose we wish to blend cattle feed from oats and barley. We are interested in five nutritional factors that are contained in these grains. The amount of these factors ($r, s, t, u,$ and v) contained in 1 bushel of each grain is shown in the following chart, together with the total amount needed.

	r	s	t	u	v
oats	3	2	5	1	2
barley	2	4	1	3	1
needed	28	30	22	20	25

EXERCISES

The costs per bushel are $.50 for oats and $.40 for barley. How many bushels of oats and barley should be blended to meet the nutritional requirements at the lowest cost?

27.6. A certain perfume manufacturer makes four kinds of perfume, which he calls Musty, Dusty, Crusty, and Rusty. Each bottle of Musty is mixed from 2 ounces of olive oil, 1 ounce of lemon juice, and 1 ounce of vanilla extract. Each bottle of Dusty is mixed from 3 ounces of olive oil, 1 ounce of lemon juice, and 2 ounces of vanilla extract. Each bottle of Crusty is mixed from 1 ounce of olive oil and 3 ounces of vanilla extract. Each bottle of Rusty is mixed from 1 ounce of olive oil, 3 ounces of lemon juice, and 2 ounces of vanilla extract. He makes a profit of $8 on each bottle of Musty he sells, $2 on each bottle of Dusty, $6 on each bottle of Crusty, and $5 on each bottle of Rusty. If he has 120 ounces of olive oil on hand, 240 ounces of lemon juice, and 310 ounces of vanilla extract, and if he can sell whatever he makes, how much of each kind of perfume should he make in order to maximize his profit?

27.7. Bob wishes to invest $12,000 in mutual funds, bonds, and a savings account. Because of the risk involved in mutual funds, he wants to invest no more in mutual funds than the sum of his bonds and savings. He also wants at least half as much in savings as he has in bonds. His expected return on the mutual funds is 11 percent, on the bonds is 8 percent, and on savings is 6 percent. How much money should Bob invest in each area in order to have the largest return on his investment?

27.8. Suppose in Exercise 27.7 that the return on bonds drops to 5 percent. If his goals remain the same, how will this effect the optimal distribution of his investment funds?

27.9. Bill wishes to invest $35,000 in stocks, bonds, and gold coins. He knows that his rate of return will depend on the economic climate of the country, which is, of course, difficult to predict. After careful analysis he determines the profit (in thousands of dollars) that he would expect on each type of investment depending on whether the economy is strong, stable, or weak:

	strong	stable	weak
stocks	4	1	-2
bonds	1	3	0
gold	-1	0	4

How should Bill invest his money in order to maximize his profit regardless of what the economy does? That is, consider the problem as a two-person matrix game in which Bill is playing against "fate."

11
CONVEX FUNCTIONS

Throughout this book we have looked at convex sets and their applications primarily from a geometrical point of view. Many of the same topics can also be developed from a functional point of view, as has been done by Rockafellar (1970) and Roberts and Varberg (1973). In this final chapter we will present an introduction to the theory of convex functions, showing how they relate to convex sets and deriving some properties that are of interest in their own right. One of the important applications of convex functions is in the study of optimization, wherein the objective function and linear constraints of a linear programming problem are replaced by convex functions. The development of the theory and techniques of convex programming is beyond the scope of this book, and the reader is referred to Stoer and Witzgall (1970), and Goldstein (1972), as well as the books mentioned previously.

SECTION 28. BASIC PROPERTIES

The definition of a convex function can be given either analytically or geometrically. We begin with the analytical definition and then show how it is related to the geometry of convex sets.

28.1. Definition. Let f be a real-valued function defined on a convex subset D of E^n. Then f is **convex** if

$$f(\alpha x + \beta y) \leq \alpha f(x) + \beta f(y) \tag{1}$$

for all x and y in D and for all $\alpha \geq 0, \beta \geq 0$ with $\alpha + \beta = 1$. A **concave** function is a function whose negative is convex. Equivalently, f is concave if it satisfies (1) with the inequality reversed. (See Exercise 28.1.)

The graph of a convex function is a subset of the product space $E^n \times R \equiv (E^n, R) = E^{n+1}$. Although the graph itself is not generally a convex set, it is the "lower boundary" of a convex set in the following sense.

BASIC PROPERTIES

28.2. Definition. Let f be a real-valued function defined on a convex subset D of $\mathbf{E}^n$. The **epigraph** of f, denoted by epi f, is the subset of $(\mathbf{E}^n, \mathbf{R})$ given by

$$\text{epi } f = \{(x, \mu): x \in D \text{ and } \mu \geq f(x)\}.$$

28.3. Theorem. Let f be a real-valued function defined on a convex subset D of $\mathbf{E}^n$. Then f is convex iff epi f is a convex subset of $(\mathbf{E}^n, \mathbf{R})$.

PROOF. Suppose f is convex and let (x_1, μ_1) and (x_2, μ_2) be points in epi f. Then for $\alpha \geq 0$, $\beta \geq 0$ and $\alpha + \beta = 1$, we have

$$\alpha\mu_1 + \beta\mu_2 \geq \alpha f(x_1) + \beta f(x_2) \geq f(\alpha x_1 + \beta x_2).$$

Therefore, the point

$$\alpha(x_1, \mu_1) + \beta(x_2, \mu_2) = (\alpha x_1 + \beta x_2, \alpha\mu_1 + \beta\mu_2)$$

is in epi f and epi f is convex.

Conversely, suppose epi f is a convex subset of $(\mathbf{E}^n, \mathbf{R})$ and let x_1 and x_2 be points in D. Then (x_1, μ_1) and (x_2, μ_2) are in epi f, where $\mu_i = f(x_i)$, $i = 1, 2$. Since epi f is convex, we have

$$\alpha(x_1, \mu_1) + \beta(x_2, \mu_2) = (\alpha x_1 + \beta x_2, \alpha\mu_1 + \beta\mu_2)$$

is in epi f for all $\alpha \geq 0$, $\beta \geq 0$, $\alpha + \beta = 1$. But this implies

$$f(\alpha x_1 + \beta x_2) \leq \alpha\mu_1 + \beta\mu_2 = \alpha f(x_1) + \beta f(x_2),$$

so that f is a convex function. ∎

It is readily apparent (and will be shown analytically) that $f(x) = x^2$ is a convex function on $\mathbf{E}^1$. It follows that its epigraph is a convex subset of $\mathbf{E}^2$. (See Figure 28.1.)

Not only does a convex function have a convex epigraph in $(\mathbf{E}^n, \mathbf{R})$, but convex subsets of $(\mathbf{E}^n, \mathbf{R})$ can be used to generate convex functions. Intuitively,

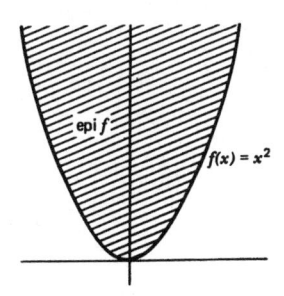

Figure 28.1.

if we are given a convex set F in (E^n, R), we want to define our function so that its graph is the "lower boundary" of F. If we require F to be "pointwise bounded below" so that it has a lower boundary, then this will be possible, as shown in the following theorem.

28.4. Theorem. Let F be a convex subset of (E^n, R) and let D be the projection of F on E^n. Suppose that $\inf\{\mu: (x, \mu) \in F\} > -\infty$ for each x in D. Define $f: D \to R$ by

$$f(x) = \inf\{\mu: (x, \mu) \in F\}.$$

Then f is a convex function on D.

PROOF. Let $x, y \in D$ and let $\alpha \geq 0, \beta \geq 0$ with $\alpha + \beta = 1$. Then $(x, f(x))$ and $(y, f(y))$ are in cl F. Thus

$$\alpha(x, f(x)) + \beta(y, f(y)) = (\alpha x + \beta y, \alpha f(x) + \beta f(y))$$

is in cl F since cl F is convex. It follows that

$$f(\alpha x + \beta y) = \inf\{\mu: (\alpha x + \beta y, \mu) \in F\}$$

$$= \inf\{\mu: (\alpha x + \beta y, \mu) \in \text{cl } F\}$$

$$\leq \alpha f(x) + \beta f(y).$$

Hence f is convex on D. ∎

In order to get a better intuitive feeling for convex functions, let us digress for a moment from our general discussion in E^n and look at the simpler case of a real-valued function f defined on an open interval (a, b) in R. Suppose that x and y are points in (a, b) with $x < y$ and that $z \equiv \alpha x + \beta y$ is any convex combination of x and y. Let x', y' and z' be the points on the graph of f corresponding to $x, y,$ and z, respectively. (See Figure 28.2.) Then the convex-

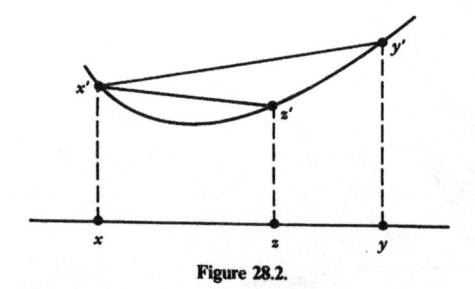

Figure 28.2.

BASIC PROPERTIES

ity of f is equivalent to requiring that z' be on or below the chord $\overline{x'y'}$. In terms of slopes, this means that

$$\text{slope } \overline{x'z'} \leq \text{slope } \overline{x'y'} \leq \text{slope } \overline{z'y'}.$$

This leads us to suspect that the derivative of f (if it exists) must be a nondecreasing function and that f'' must be nonnegative. Our intuition is justified by the following theorem.

28.5. Theorem. Let f be a real-valued function defined on the open interval (a, b) in R, and suppose that f'' exists on (a, b). Then f is convex iff $f''(t) \geq 0$ for all t in (a, b).

PROOF. Suppose f is convex and consider points x and y in (a, b) with $x < y$. Let $t_1 \equiv y, t_2, t_3, \ldots$ be a decreasing sequence of points converging to x, and let $t_1' \equiv y', t_2', t_3', \ldots$ be the corresponding points on the graph of f. (See Figure 28.3.) It follows that

$$f'(x) = \lim_{k \to \infty} \frac{f(x) - f(t_k)}{x - t_k}$$

$$= \lim_{k \to \infty} \text{slope } \overline{x't_k'} = \lim_{k \to \infty} m_k,$$

where m_k is the slope of the chord $\overline{x't_k'}$ ($k = 1, 2, \ldots$). But from the discussion preceding the theorem, m_k is a decreasing sequence. Thus $f'(x) \leq m_k$ for each k, and in particular,

$$f'(x) \leq m_1 = \text{slope } \overline{x'y'}.$$

A similar argument shows that the slope of $\overline{x'y'}$ is less than or equal to $f'(y)$, so that $f'(x) \leq f'(y)$. It follows that f' is nondecreasing and $f'' \geq 0$ on (a, b).

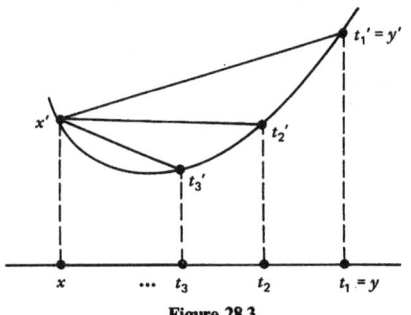

Figure 28.3.

Conversely, suppose that $f'' \geq 0$ on (a, b) so that f' is nondecreasing on (a, b). Let x and y be points in (a, b) with $x < y$ and let $z \equiv \alpha x + \beta y$ be a convex combination of x and y. From the fundamental theorem of integral calculus, we have

$$f(z) - f(x) = \int_x^z f'(t)\, dt \leq f'(z)(z - x),$$

where the inequality holds because the maximum of $f'(t)$ on (x, z) is attained at z. Similarly,

$$f(y) - f(z) = \int_z^y f'(t)\, dt \geq f'(z)(y - z)$$

since the minimum of $f'(t)$ on (z, y) is attained at z. Thus

$$f(z) \leq f(x) + f'(z)(z - x)$$

and

$$f(z) \leq f(y) - f'(z)(y - z).$$

It follows that

$$f(z) = \alpha f(z) + \beta f(z)$$
$$\leq \alpha [f(x) + f'(z)(z - x)] + \beta [f(y) - f'(z)(y - z)]$$
$$= \alpha f(x) + \beta f(y) + f'(z)[\alpha(z - x) - \beta(y - z)]$$
$$= \alpha f(x) + \beta f(y),$$

since $\alpha(z - x) - \beta(y - z) = (\alpha + \beta)z - (\alpha x + \beta y) = z - z = 0$. Thus f is a convex function. ∎

By using Theorem 28.5 we can readily identify many common functions as being convex, at least on appropriately restricted domains. For example, each of the following is convex:

x^{2k} on $(-\infty, \infty)$ for $k = 1, 2, 3, \ldots$
x^{2k+1} on $(0, \infty)$ for $k = 1, 2, 3, \ldots$
$\sin x$ on $(\pi, 2\pi)$
$e^{\alpha x}$ on $(-\infty, \infty)$ for $\alpha \in \mathbf{R}$
$-\ln x$ on $(0, \infty)$

We now return to the general context of functions defined on E^n and derive an important inequality that is sometimes taken as the definition of a convex function.

BASIC PROPERTIES

28.6. Theorem (Jensen's Inequality). Let f be a real-valued function defined on a convex subset D of E^n, and let $\alpha_1 x_1 + \cdots + \alpha_m x_m$ be a convex combination of points $x_1, \ldots, x_m$ in D. That is, $\alpha_i \geq 0$ ($i = 1, \ldots, m$) and $\alpha_1 + \cdots + \alpha_m = 1$. Then f is convex iff

$$f(\alpha_1 x_1 + \cdots + \alpha_m x_m) \leq \alpha_1 f(x_1) + \cdots + \alpha_m f(x_m).$$

PROOF. Suppose f is a convex function. The proof of the inequality is by induction on the number m of points in the convex combination. If $m = 2$, the inequality corresponds to the one in Definition 28.1. We now suppose that $m > 2$ and that the inequality holds for any collection of $m - 1$ or fewer points in D. If we let $\beta = \alpha_2 + \cdots + \alpha_m$, then $(\alpha_2/\beta) x_2 + \cdots + (\alpha_m/\beta) x_m$ is a convex combination of the $m - 1$ points $x_2, \ldots, x_m$. But $\alpha_1 + \beta = 1$ and $\beta \geq 0$, so we conclude that

$$\begin{aligned} f(\alpha_1 x_1 + \cdots + \alpha_m x_m) &= f\left[\alpha_1 x_1 + \beta\left(\frac{\alpha_2}{\beta} x_2 + \cdots + \frac{\alpha_m}{\beta} x_m\right)\right] \\ &\leq \alpha_1 f(x_1) + \beta f\left(\frac{\alpha_2}{\beta} x_2 + \cdots + \frac{\alpha_m}{\beta} x_m\right) \\ &\leq \alpha_1 f(x_1) + \beta\left[\frac{\alpha_2}{\beta} f(x_2) + \cdots + \frac{\alpha_m}{\beta} f(x_m)\right] \\ &= \alpha_1 f(x_1) + \alpha_2 f(x_2) + \cdots + \alpha_m f(x_m). \end{aligned}$$

The converse is immediate. ∎

We conclude this section with two theorems which show that the convexity property of a function is preserved under certain kinds of compositions and linear combinations. Using them, we can obtain more complicated convex functions by appropriately combining simpler functions that are already known to be convex.

28.7. Theorem. Let f be a convex function defined on a convex subset D of E^n, and let g be a convex function defined on an interval I in R. Suppose that $f(D) \subset I$ and that g is nondecreasing. Then the composition $g \circ f$ is convex on D.

PROOF. For x and y in D and $\alpha \geq 0, \beta \geq 0$ with $\alpha + \beta = 1$, we have $f(\alpha x + \beta y) \leq \alpha f(x) + \beta f(y)$, since f is convex. Since g is nondecreasing and convex, we then obtain

$$\begin{aligned} g(f(\alpha x + \beta y)) &\leq g(\alpha f(x) + \beta f(y)) \\ &\leq \alpha g(f(x)) + \beta g(f(y)). \quad \blacksquare \end{aligned}$$

28.8. Theorem. Let f and g be convex functions defined on the convex subset D of E^n and let $\delta > 0$. Then $f + g$ and δf are convex on D.

PROOF. Exercise 28.3. ∎

EXERCISES

28.1. Let f be a real-valued function defined on a convex subset D of E^n. Prove that f is concave iff $f(\alpha x + \beta y) \geq \alpha f(x) + \beta f(y)$ for all $x, y \in D$ and all $\alpha \geq 0, \beta \geq 0$ with $\alpha + \beta = 1$.

28.2. Prove that the Euclidean norm function $f(x) \equiv \|x\|$ is convex on E^n. In particular, for $x \in \mathsf{R}$ we have $f(x) \equiv |x|$ is convex on R.

28.3. Prove Theorem 28.8.

28.4. Let Q be a convex quadrilateral in E^2. Construct a square on each of the sides of Q. (See Figure 28.4.) Show that if the perimeter of the quadrilateral is not less than 6, then the sum of the areas of the four squares is at least 9.

28.5. The sum of the cubes of four positive numbers is at most 108. Show that the sum of the original four numbers cannot exceed 12.

28.6. Let $D \equiv \{(\alpha, \beta): \alpha \geq 0 \text{ and } \beta \geq 0\}$. Prove or find a counterexample:
(a) The function $f[(\alpha, \beta)] = (\alpha^2 + 2\beta^3)^2$ is convex on D.
(b) The function $f[(\alpha, \beta)] = 2\alpha\beta$ is convex on D.

28.7. Let $f_1, \ldots, f_m$ be convex functions defined on a convex set D, and let g be the function defined by $g(x) = \max\{f_i(x): i = 1, \ldots, m\}$.
(a) Prove that g is convex on D by using Definition 28.1.
(b) Prove that g is convex on D by showing that epi $g = \bigcap_{i=1}^m$ epi f_i.

28.8. Prove that $2e^{x+y} \leq e^{2x} + e^{2y}$ for all real x and y.

28.9. Prove that $(x_1 \cdots x_k)^{1/k} \leq (x_1 + \cdots + x_k)/k$ for positive real numbers x_i ($i = 1, \ldots, k$). Thus the geometric mean is always less than or equal to the arithmetic mean.

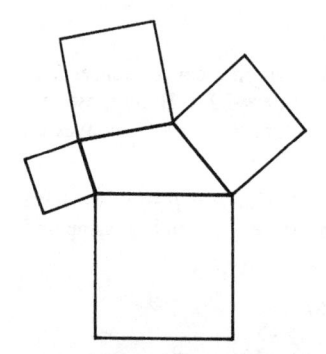

Figure 28.4.

28.10. Let f be a convex function defined on a polytope P in E^n. Prove that $\sup_{x \in P} f(x) = f(x_0)$ for some extreme point x_0 in P.

28.11. Let δ be a positive integer. A real-valued function f defined on a convex cone C with vertex at θ is said to be **homogeneous of degree δ** if $f(\lambda x) = \lambda^\delta f(x)$ for any $x \in C$ and any $\lambda \geq 0$. Suppose that f is a convex function defined on C and that $f(x) > 0$ for all $x \neq \theta$ in C. If f is homogeneous of degree δ, prove that the function g defined on C by $g(x) \equiv [f(x)]^{1/\delta}$ is convex on C.

28.12. Let $\alpha_1, \ldots, \alpha_m$ and $\beta_1, \ldots, \beta_m$ be nonnegative real numbers and let p be a positive integer. Prove the *Minkowski inequality*:

$$\left[\sum_{i=1}^{m}(\alpha_i + \beta_i)^p\right]^{1/p} \leq \left[\sum_{i=1}^{m}\alpha_i^p\right]^{1/p} + \left[\sum_{i=1}^{m}\beta_i^p\right]^{1/p}$$

28.13. Which number is larger: $(3^4 + 7^4 + 11^4)^{1/4}$ or $(1^4 + 3^4 + 5^4)^{1/4} + (2^4 + 4^4 + 6^4)^{1/4}$?

28.14. Prove the following: If f is a convex function then for every real number δ, the level set $L_\delta \equiv \{x : f(x) \leq \delta\}$ is convex.

28.15. Find an example to show that the converse of Exercise 28.6 is not true. That is, find a nonconvex function such that its level sets are all convex.

28.16. Let f be a real-valued function defined on a convex set D. Then f is said to be **quasiconvex** if the level sets $L_\delta \equiv \{x : f(x) \leq \delta\}$ are convex for each $\delta \in \mathsf{R}$. Prove that f is quasiconvex iff for all $x, y \in D$ and $\alpha, \beta \geq 0$ with $\alpha + \beta = 1$, we have $f(\alpha x + \beta y) \leq \max\{f(x), f(y)\}$.

*28.17. Let f be a real-valued function defined on a convex set D. Then f is said to be **mid-convex** if $f[(x + y)/2] \leq \frac{1}{2}[f(x) + f(y)]$ for all $x, y \in D$. Prove the following:
(a) f is mid-convex iff for any points $x_1, \ldots, x_m$ in D and for any *rational* numbers $\alpha_1, \ldots, \alpha_m$ such that each $\alpha_i \geq 0$ and $\sum_{i=1}^{m}\alpha_i = 1$, we have $f(\sum_{i=1}^{m}\alpha_i x_i) \leq \sum_{i=1}^{m}\alpha_i f(x_i)$.
(b) Let f be a mid-convex function defined on an open convex set D. If f is continuous, then f is convex.

SECTION 29. SUPPORT AND DISTANCE FUNCTIONS

Although convex functions arise in many different settings and applications, two particular convex functions are closely related to the geometry of convex sets: support functions and distance functions. In this section we derive some of their basic properties and establish the dual relationship between them.

29.1. Definition. Let S be a nonempty convex set. The **support function** h of S is the real-valued function defined by

$$h(x) \equiv \sup_{s \in S} \langle s, x \rangle$$

for all x for which the supremum is finite.

29.2. Theorem. Let S be a nonempty convex set, let h be the support function of S, and let D be the domain of h. Then the following hold:
   (a) The domain D is a convex cone with vertex at the origin.
   (b) The function h is convex.
   (c) The function h is **positively homogeneous**: $h(\lambda x) = \lambda h(x)$ for $\lambda \geq 0$ and $x \in D$.

PROOF. Let $x \in D$. Then for $\lambda \geq 0$, we have

$$h(\lambda x) \equiv \sup_{s \in S} \langle s, \lambda x \rangle = \lambda \sup_{s \in S} \langle s, x \rangle = \lambda h(x) < \infty.$$

Thus $\lambda x \in D$ so that D is a cone with vertex at the origin. Since $h(\lambda x) = \lambda h(x)$, h is positively homogeneous.

Now let $x \in D$ and $y \in D$. Then for $\alpha \geq 0$ and $\beta \geq 0$ with $\alpha + \beta = 1$ we have

$$h(\alpha x + \beta y) = \sup_{s \in S} \langle s, \alpha x + \beta y \rangle = \sup_{s \in S} \{\alpha \langle s, x \rangle + \beta \langle s, y \rangle\}$$

$$\leq \sup_{s \in S} \{\alpha \langle s, x \rangle\} + \sup_{s \in S} \{\beta \langle s, y \rangle\}$$

$$= \alpha \sup_{s \in S} \langle s, x \rangle + \beta \sup_{s \in S} \langle s, y \rangle$$

$$= \alpha h(x) + \beta h(y) < \infty.$$

It follows that $\alpha x + \beta y \in D$ so that D is convex. Furthermore, we see that the function h is convex. ∎

The name "support" function is justified by the following theorem.

29.3. Theorem. Let S be a nonempty compact convex set and let h be its support function. If u is any fixed point other than the origin, then the following hold:
   (a) There exists a point x_u in S such that $h(u) = \langle x_u, u \rangle$.
   (b) The hyperplane $H \equiv \{x: \langle x, u \rangle = h(u)\}$ supports S at x_u.
   (c) The distance from H to θ is equal to $h(u/\|u\|)$.

SUPPORT AND DISTANCE FUNCTIONS

PROOF. (a) Since the inner product is continuous and S is compact, Theorem 1.23 implies the existence of x_u.

(b) The hyperplane H bounds S since, for all s in S,

$$\langle s, u \rangle \leq h(u) \equiv \sup_{s \in S} \langle s, u \rangle.$$

Since $x_u \in H \cap S$, H supports S at x_u.

(c) Since u is orthogonal to H (Exercise 3.1), there exists a real λ such that $\lambda u \in H$. But then $\|\lambda u\|$ is the distance from H to θ. Since $\lambda u \in H$, $\langle \lambda u, u \rangle = h(u)$. Thus

$$h\left(\frac{u}{\|u\|}\right) = \frac{1}{\|u\|} h(u) = \frac{\langle \lambda u, u \rangle}{\|u\|} = |\lambda| \frac{\langle u, u \rangle}{\|u\|}$$

$$= |\lambda| \|u\| = \|\lambda u\|. \blacksquare$$

29.4. Examples

(a) Let S be the closed unit ball centered at θ in E^n. (See Figure 29.1.) Then the support function h of S is defined for all x in E^n. For $u \neq \theta$ we have

$$h(u) = \sup_{s \in S} \langle s, u \rangle = \left\langle \frac{u}{\|u\|}, u \right\rangle = \frac{\langle u, u \rangle}{\|u\|} = \|u\|.$$

Since $h(\theta) = 0$, we conclude that for all x, $h(x) = \|x\|$. Thus the support function for the unit ball corresponds to the norm.

(b) Let T be the square in E^2 with vertices at $(0, \pm 1)$ and $(\pm 1, 0)$. (See Figure 29.2.) For $u \equiv (2, 1)$ we wish to maximize the linear form $\langle t, u \rangle$ where $t \in T$. We know that this maximum occurs at an

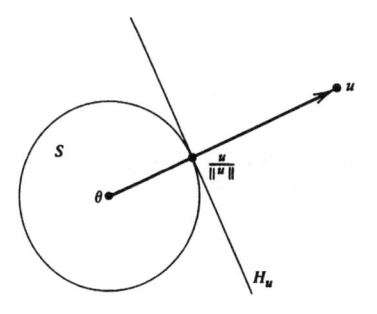

Figure 29.1.

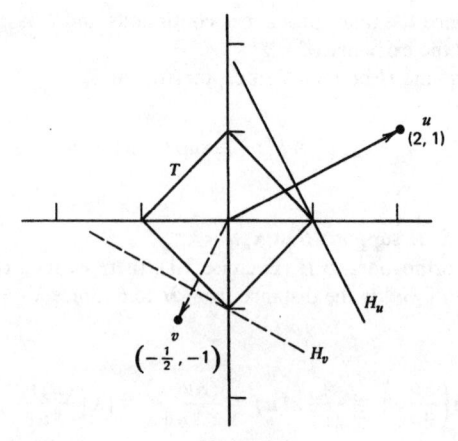

Figure 29.2.

extreme point of T (Theorem 5.7), and this is easily seen to be the vertex $(1, 0)$. We find

$$h(u) = \langle (1,0), (2,1) \rangle = 2 + 0 = 2.$$

Geometrically, $(1, 0)$ is the point of T that is "farthest" in the direction of u. If we take $v \equiv (-\frac{1}{2}, -1)$, then the corresponding supporting hyperplane H_v passes through $(0, -1)$, and we have

$$h(v) = \langle (0, -1), (-\tfrac{1}{2}, -1) \rangle = 0 + 1 = 1.$$

Since T is bounded, the domain of h is all of E^2. In general we have

$$h[(x, y)] = \max\{|x|, |y|\}$$

for (x, y) in E^2.

(c) Let S be the region in E^2 defined by

$$S \equiv \{(x, y): 0 \leq y \leq x - 1\}.$$

(See Figure 29.3.) Since S is unbounded, the domain D of the support function h of S is a proper subset of E^2, and we find

$$D = \{(x, y): x \leq 0 \text{ and } y \leq -x\}.$$

SUPPORT AND DISTANCE FUNCTIONS

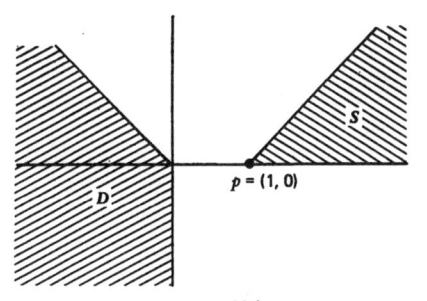

Figure 29.3.

Since S has only one extreme point, namely $p \equiv (1, 0)$, the support function h simplifies to

$$h[(x, y)] = \langle (1,0), (x, y) \rangle = x,$$

where $(x, y) \in D$.

Since the support function of a convex set S can be used to obtain the hyperplanes that support S, we might expect that these supporting hyperplanes could in turn be used to generate the convex set S itself. We now show that this is in fact the case.

29.5. Theorem. Let S be a nonempty closed convex set and let h be its support function with a nonempty domain D. Then S can be characterized by the condition

$$S = \{x : \langle x, u \rangle \leq h(u) \text{ for all } u \in D\},$$

or, equivalently,

$$S = \bigcap_{u \in D} \{x : \langle x, u \rangle \leq h(u), u \text{ fixed in } D\}.$$

PROOF. The set

$$H_u \equiv \{x : \langle x, u \rangle \leq h(u), u \text{ fixed in } D\}$$

is a closed half-space containing S, for each $u \in D$. It follows that $S \subset \bigcap_{u \in D} H_u$.

On the other hand, suppose $x_0 \notin S$. (See Figure 29.4.) Theorem 4.12 implies the existence of a hyperplane H strictly separating x_0 and S, and we may write $H = \{x : \langle x, u_0 \rangle = \delta\}$ for some point $u_0 \neq \theta$ and some real number δ. (See Corollary 3.5.) Without loss of generality we may assume that $\langle s, u_0 \rangle < \delta$ for

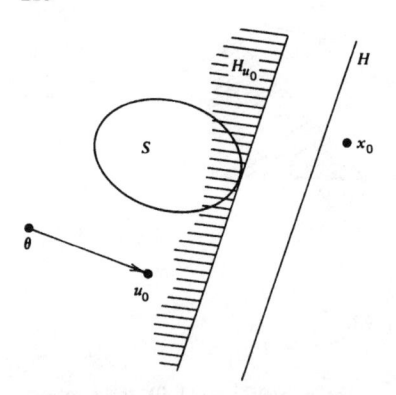

Figure 29.4.

all $s \in S$ and that $\langle x_0, u_0 \rangle > \delta$. It follows that

$$\sup_{s \in S} \langle s, u_0 \rangle \leq \delta < \infty,$$

so that $u_0 \in D$. But then $h(u_0) \leq \delta$, so that $x_0 \notin H_{u_0}$. Hence $x_0 \notin \bigcap_{u \in D} H_u$. Combining the preceding arguments we obtain $S = \bigcap_{u \in D} H_u$. ∎

Although the support function was defined for any nonempty convex set, the distance function will be defined only for convex bodies that contain the origin in their interiors.

29.6. Definition. Let S be a compact convex body with $\theta \in \text{int } S$. Then the **Minkowski distance function** p of S is the real-valued function defined by

$$p(x) \equiv \inf\{\lambda : \lambda > 0 \text{ and } x \in \lambda S\},$$

for all x in $\mathbf{E}^n$.

This distance function was first defined by Minkowski in 1911. It provides a useful method of obtaining a norm (and hence a topology) in very general finite-dimensional (and infinite-dimensional) linear spaces. In n-dimensional Euclidean space we can visualize $p(x)$ as follows: For $x \neq \theta$, let x_0 be the unique point at which the ray R from θ through x intersects bd S. (See Figure 29.5.) Then $x = p(x)x_0$ and $p(x) = \|x\|/\|x_0\|$. We refer to p as the *distance* function of S because if we think of the Euclidean distance from θ to x_0 as one unit of "distance relative to S" in the direction of R, then $p(x)$ is the "distance relative to S" from θ to x. If S is not a Euclidean ball centered at θ, then the basic unit of distance relative to S depends on the direction. We encountered this same phenomenon in Example 19.3 when discussing nearest-point properties. We now show how our earlier example of a non-Euclidean distance can be derived from a Minkowski distance function.

SUPPORT AND DISTANCE FUNCTIONS

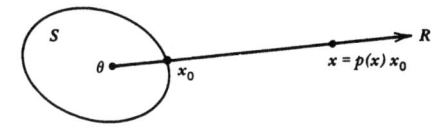

Figure 29.5.

29.7. Example. Let S be the square in E^2 with vertices at $(1, \pm 1)$ and $(-1, \pm 1)$. (See Figure 29.6.) If the ray from θ through the point (x, y) intersects a vertical side of the square S, then $p[(x, y)]$ will equal $|x|$. For example, we see that $p[(3, 1)] = 3$. On the other hand, if the ray from θ through (x, y) intersects a horizontal side of S, then $p[(x, y)] = |y|$. Thus $p[(-\frac{1}{2}, -2)] = 2$. In general we find $p[(x, y)] = \max\{|x|, |y|\}$.

If we define the "d_2-distance" between two points (x_1, y_1) and (x_2, y_2) to be p of their difference, then we obtain

$$d_2[(x_1, y_1), (x_2, y_2)] = p[(x_1 - x_2, y_1 - y_2)]$$
$$= \max\{|x_1 - x_2|, |y_1 - y_2|\},$$

as in Example 19.3.

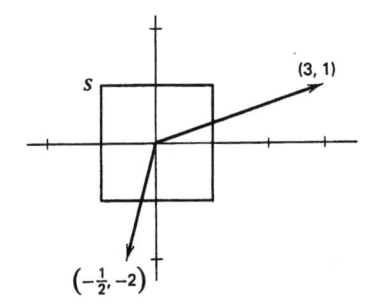

Figure 29.6.

In comparing Examples 29.7 and 29.4(b), we see that the support function of $T \equiv \operatorname{conv}\{(0, 1), (0, -1), (1, 0), (-1, 0)\}$ is the same as the distance function of $S \equiv \operatorname{conv}\{(1, 1), (1, -1), (-1, 1), (-1, -1)\}$. Since T is the polar set of S, this illustrates the following general theorem.

29.8. Theorem. Let S be a compact convex body in E^n with $\theta \in \operatorname{int} S$ and let S^* be the polar set of S. Then the support function of S is equal to the Minkowski distance function of S^*, and the Minkowski distance function of S is equal to the support function of S^*. That is, $h_S = p_{S^*}$ and $p_S = h_{S^*}$.

PROOF. From Theorem 23.4 we know that S^* is a compact convex set containing the origin in its interior. Thus for $x \neq \theta$, we may let x_0 be the point

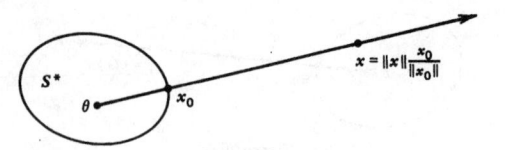

Figure 29.7.

where the ray from θ through x intersects bd S^*. (See Figure 29.7.) Since $x_0 \in S^*$ we have $\langle s, x_0 \rangle \leq 1$ for all $s \in S$, so that $h_S(x_0) = \sup_{s \in S} \langle s, x_0 \rangle \leq 1$. Since $x = \|x\|(x_0/\|x_0\|)$, it follows that

$$h_S(x) = h_S\left(\frac{\|x\|}{\|x_0\|} x_0\right) = \frac{\|x\|}{\|x_0\|} h_S(x_0)$$

$$\leq \frac{\|x\|}{\|x_0\|} = p_{S^*}(x)$$

On the other hand, $p_S(x) > 0$ for $x \neq \theta$, and so we have $x \notin \delta S$ for every δ such that $0 < \delta < p_S(x)$. But $\delta S = (\delta S)^{**} = [(1/\delta)S^*]^*$ by Theorems 23.3 and 23.5, so $x \notin [(1/\delta)S^*]^*$. This means that there exists a point $(1/\delta)s_0^*$ in $(1/\delta)S^*$ such that $\langle (1/\delta)s_0^*, x \rangle > 1$, or equivalently, $\langle s_0^*, x \rangle > \delta$. It follows that

$$h_{S^*}(x) = \sup_{s^* \in S^*} \langle s^*, x \rangle > \delta.$$

Since this holds for each $\delta < p_S(x)$, we conclude that $h_{S^*}(x) \geq p_S(x)$.

Finally, since $h_{S^*}(\theta) = 0 = p_S(\theta)$, we have $h_{S^*}(x) = p_S(x)$ for all x in E^n. Replacing S by S^*, we obtain

$$h_S(x) = h_{S^{**}}(x) = p_{S^*}(x)$$

for all x, since $S = S^{**}$. ∎

The dual relationship between support functions and distance functions exhibited in the preceding theorem implies that for compact S with $\theta \in \text{int } S$, the Minkowski distance function $p(x)$ is a convex function defined for all x in E^n which is positively homogeneous. From this it is easy to derive (Exercise 29.1) that $p(x)$ is **subadditive**. That is, for all x and y in E^n,

$$p(x + y) \leq p(x) + p(y).$$

These properties make $p(x)$ useful in defining a norm in an arbitrary linear space. (See Exercise 29.5.)

EXERCISES

29.1. Let f be a real-valued function defined on a convex cone D with vertex at the origin. Suppose that f is convex and positively homogeneous. Prove that f is subadditive [i.e., prove that for every $x, y \in D$, $f(x + y) \leq f(x) + f(y)$].

29.2. Let $A \equiv \{(\alpha_1, \ldots, \alpha_n): \alpha_i \geq 0, \alpha_1 + \cdots + \alpha_n = 1\}$ and let h be the support function of A. Show that $h(x) = \max\{\xi_i: i = 1, \ldots, n\}$ where $x \equiv (\xi_1, \ldots, \xi_n)$.

29.3. Let S be a nonempty convex set with support function h defined on D. Prove that the dual cone of S corresponds to the epigraph of h. That is, define

$$C \equiv \{(\alpha, x): x \in D \text{ and } \alpha \geq h(x)\}$$

and prove that $C = \text{dc } S$.

29.4. Let S be a nonempty compact convex subset of $\mathbf{E}^2$ and let h be its support function. If u is a unit vector parallel to a given line ℓ, then prove that the width of S in the direction of ℓ is equal to $h(u) + h(-u)$.

29.5. Let S be a compact convex set with $\theta \in \text{int } S$ and suppose that S is **centrally symmetric with respect to θ**. That is, $-S = S$. Prove that the Minkowski distance function $p(x)$ of S satisfies the properties of a norm by showing that
(a) $p(x) \geq 0$ for all x.
(b) $p(x) > 0$ if $x \neq \theta$.
(c) $p(\alpha x) = |\alpha| p(x)$ for all x and all real α.
(d) $p(x + y) \leq p(x) + p(y)$ for all x and y.

29.6. Let A and B be nonempty convex subsets of $\mathbf{E}^n$. Let h_S denote the support function of any convex set S, and denote the domain of h_S by D_S. Prove that $h_{A+B} = h_A + h_B$ and that $D_{A+B} = D_A \cap D_B$.

29.7. Let A and B be nonempty compact convex sets with support functions h_A and h_B, respectively. Prove that $A \subset B$ iff $h_A(x) \leq h_B(x)$ for all x in $\mathbf{E}^n$.

29.8. Let A and B be compact convex bodies both of which contain the origin as an interior point. Let h_S (p_S, respectively) denote the support (Minkowski distance, respectively) function of any convex set S. Prove the following:
(a) $h_{u+A}(x) = h_A(x) + \langle u, x \rangle$ for all $x \in \mathbf{E}^n$, $u \in \mathbf{E}^n$.
(b) $h_C(x) = \max\{h_A(x), h_B(x)\}$ for all $x \in \mathbf{E}^n$, where $C \equiv \text{conv}(A \cup B)$.
(c) $p_{A \cap B}(x) = \max\{p_A(x), p_B(x)\}$ for all $x \in \mathbf{E}^n$.

29.9. Let S be a compact subset of $\mathbf{E}^n$ which is star-shaped relative to θ and which contains θ in its interior. Let q be the real-valued function defined by

$$q(x) \equiv \inf\{\lambda : \lambda > 0 \text{ and } x \in \lambda S\}$$

for all $x \in E^n$. Prove that S is convex iff q is subadditive and positively homogeneous, that is, iff

$$q(x+y) \leq q(x) + q(y) \qquad \text{for all } x, y \in E^n$$

$$q(\alpha x) = \alpha q(x) \qquad \text{for all } \alpha \geq 0 \text{ and } x \in E^n.$$

29.10. Let f be a real-valued function defined on E^n which is subadditive and positively homogeneous. (Such a function is called a **gauge function**.)
(a) Prove that f is convex.
(b) A vector u is said to be a **linearity vector** of the gauge function f if $f(\lambda u) = \lambda f(u)$ for all real λ. Suppose u_1 and u_2 are linearity vectors of the gauge function f. Prove that $\alpha u_1 + \beta u_2$ is also a linearity vector for f where α and β are any real numbers.

29.11. Let g be a real-valued function defined on E^n that satisfies:

$$g(x+y) \leq g(x) + g(y) \qquad \text{for all } x, y \in E^n$$

$$g(\alpha x) = |\alpha| g(x) \qquad \text{for all } x \in E^n \text{ and } \alpha \in R.$$

(Such a function is called a **semi-norm** on E^n.) Prove the following:
(a) g is convex.
(b) $g(\theta) = 0$.
(c) $g(x) \geq 0$ for all $x \in E^n$.
(d) $|g(x) - g(y)| \leq g(x-y)$ for all $x, y \in E^n$.

SECTION 30. CONTINUITY AND DIFFERENTIABILITY

Although the main thrust of this chapter—and indeed the whole book—has been geometrical, no discussion of convex functions would be complete without some mention of the analytic properties of continuity and differentiability. While the situation in an arbitrary (possibly infinite-dimensional) normed linear space is somewhat more complicated, within the context of E^n the requirement that a function be convex is sufficient to guarantee several strong analytic properties. We begin with boundedness and then discuss continuity.

30.1. Theorem. Let f be a convex function defined on the convex set S and let T be a nonempty compact subset of relint S. Then f is bounded above on T. That is, there exists a real number M such that $f(x) \leq M$ for all x in T.

PROOF. Theorem 22.7 implies the existence of a polytope P such that $T \subset P \subset S$. Thus it suffices to show that f is bounded above on every polytope

CONTINUITY AND DIFFERENTIABILITY

P contained in S. To this end, let $\{x_1, \ldots, x_m\}$ be the set of vertices of P. Then for every x in P we have

$$x = \sum_{i=1}^{m} \alpha_i x_i,$$

where $0 \leq \alpha_i \leq 1$ and $\sum_{i=1}^{m} \alpha_i = 1$. It follows from Jensen's inequality (Theorem 28.6) that

$$f(x) = f\left(\sum_{i=1}^{m} \alpha_i x_i\right) \leq \sum_{i=1}^{m} \alpha_i f(x_i)$$

$$\leq M \sum_{i=1}^{m} \alpha_i = M,$$

where $M = \max\{f(x_i): i = 1, \ldots, m\}$. ∎

30.2. Theorem. Let f be a convex function defined on an open convex set D. Then f is continuous at every point of D.

PROOF. Let $x \in D$ and let B be a closed ball of positive radius centered at x with $B \subset D$. Then for any u such that $x + u \in B$ and for any λ such that $0 < \lambda \leq 1$, we have $x + \lambda u \in B$. Thus

$$f(x + \lambda u) = f[\lambda(x + u) + (1 - \lambda)x]$$

$$\leq \lambda f(x + u) + (1 - \lambda)f(x),$$

and we obtain

$$f(x + \lambda u) - f(x) \leq \lambda[f(x + u) - f(x)].$$

Since B is compact, Theorem 30.1 implies the existence of a real number M such that $f(x + u) - f(x) \leq M$ for all u such that $x + u \in B$. Hence

$$f(x + \lambda u) - f(x) \leq \lambda M.$$

A similar argument (replacing $x + \lambda u$ by x and u by $-u$) implies

$$f(x) - f(x + \lambda u) \leq \lambda M.$$

Combining these last two inequalities, we have

$$|f(x + \lambda u) - f(x)| \leq \lambda M,$$

which implies that f is continuous at x. ∎

It should not be surprising that a convex function has to be continuous. Indeed for a function f defined on R, it is clear that a break in the graph of f would cause epi f to be nonconvex. (See Figure 30.1.) We might expect as well that requiring the set epi f to be convex would effect the existence of a derivative. Typical examples of pathological functions that are continuous but nowhere differentiable (e.g., Gelbaum and Olmstead, p. 38) involve a great deal of bending, which would not be allowed for a convex function. On the other hand, it is certainly possible for a convex function to fail to have a derivative at some points.

In Section 28 we saw that the slope of a differentiable convex function of one variable is a nondecreasing function. We begin our discussion of differentiability by establishing a similar result for an arbitrary convex function defined on E^n.

30.3. Theorem. Let f be a convex function defined on an open subset D of E^n. Then the quotient

$$\frac{f(x + \lambda y) - f(x)}{\lambda}$$

is a nondecreasing function of λ for $\lambda > 0$.

PROOF. Let $x \in D$ and let $\mu = \alpha\lambda$ with $0 < \alpha \leq 1$. Then since f is convex, we have

$$f(x + \alpha\lambda y) \equiv f[(1 - \alpha)x + \alpha(x + \lambda y)]$$
$$\leq (1 - \alpha)f(x) + \alpha f(x + \lambda y).$$

It follows that

$$\frac{f(x + \alpha\lambda y) - f(x)}{\alpha} \leq f(x + \lambda y) - f(x).$$

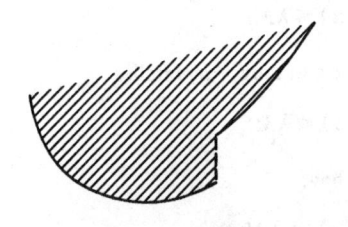

Figure 30.1.

CONTINUITY AND DIFFERENTIABILITY

Therefore,

$$\frac{f(x+\mu y)-f(x)}{\mu} = \frac{1}{\lambda}\frac{f(x+\alpha\lambda y)-f(x)}{\alpha}$$

$$\leq \frac{f(x+\lambda y)-f(x)}{\lambda}$$

for all μ such that $0 < \mu \leq \lambda$ and all λ and y such that $x + \lambda y \in D$. Since μ is an arbitrary positive number less than or equal to λ, we see that the required quotient is nondecreasing for $\lambda > 0$. ∎

Using Theorem 30.3 we can prove that a convex function has a directional derivative (also called a one-sided Gateaux differential) in each direction throughout the interior of its domain. We precede this result with a formal definition of the directional derivative.

30.4. Definition. Let f be a real-valued function defined on an open subset D of E^n. The **directional derivative** of f at x in the direction of y, denoted by $\delta f(x; y)$, is given by

$$\delta f(x; y) \equiv \lim_{\lambda \to 0+} \frac{f(x+\lambda y)-f(x)}{\lambda}$$

for all $x \in D$ and $y \in E^n$ for which the limit exists.

30.5. Theorem. Let f be a convex function defined on an open subset D of E^n. Then given any $y \in E^n$ and any $x \in D$, f has a directional derivative at x in the direction of y.

PROOF. Let $y \in E^n$ and $x \in D$. Since D is open, there exists $\alpha > 0$ such that $x + \alpha y \in D$ and $x - \alpha y \in D$. Since f is convex, we have for λ such that $0 < \lambda \leq 1$,

$$f(x) = f\left[\frac{\lambda}{1+\lambda}(x-\alpha y)+\frac{1}{1+\lambda}(x+\lambda\alpha y)\right]$$

$$\leq \frac{\lambda}{1+\lambda}f(x-\alpha y)+\frac{1}{1+\lambda}f(x+\lambda\alpha y).$$

Thus

$$(1+\lambda)f(x) \leq \lambda f(x-\alpha y)+f(x+\lambda\alpha y),$$

$$\lambda f(x) - \lambda f(x-\alpha y) \leq f(x+\lambda\alpha y)-f(x),$$

and

$$f(x) - f(x - \lambda \alpha y) \leq \frac{f(x + \lambda \alpha y) - f(x)}{\lambda}.$$

Hence the difference quotient in the definition of the directional derivative is bounded below in all directions. This together with the monotonicity property of Theorem 30.3 implies that the limit exists for all $x \in D$ and $y \in \mathsf{E}^n$. ∎

If $\delta f(x; y) = -\delta f(x; -y)$, then f is said to have two-sided directional derivative at x in the direction of y. If f has a two-sided directional derivative in all directions at x, then f is said to be **Gateaux differentiable** at x. For an arbitrary real-valued function of several variables, Gateaux differentiability is not sufficient to guarantee the existence of a unique hyperplane tangent to the graph of f. A stronger kind of differentiability, called Fréchet differentiability, is needed.

We say that f is **Fréchet differentiable** at x if there exists a vector x' such that

$$\lim_{y \to x} \frac{f(y) - f(x) - \langle x', y - x \rangle}{\|y - x\|} = 0.$$

The vector x' is called the gradient of f at x.

Fortunately, a convex function is Gateaux differentiable iff it is Fréchet differentiable, and the existence of these derivatives at a point x corresponds to the existence of a unique hyperplane of support to epi f at the point $(x, f(x))$. The proof of these assertions is beyond the scope of this book, and the reader is referred to Rockafellar (1970) or Roberts and Varberg (1973).

Thus for a convex function f defined on an open set D in E^n, there exists a (one-sided) directional derivative in each direction at each point in D. If it happens that at some point x these one-sided derivatives fit together in such a way that $\delta f(x; y) = -\delta f(x; -y)$ for all y in E^n, then f is both Gateaux and Fréchet differentiable at x, and there exists a unique hyperplane supporting epi f at $(x, f(x))$.

EXERCISES

30.1. Let f be a positive real-valued function defined on E^n that is subadditive and positively homogeneous. That is,

$$f(x) > 0 \qquad \text{for all } x \neq \theta$$

$$f(x + y) \leq f(x) + f(y) \qquad \text{for all } x, y \in \mathsf{E}^n$$

$$f(\lambda x) = \lambda f(x) \qquad \text{for all } \lambda \geq 0 \text{ and } x \in \mathsf{E}^n.$$

(Thus f is a positive gauge function.) Prove that the set $S \equiv \{x: f(x) \leq 1\}$ is a compact convex body with $\theta \in \text{int } S$. Furthermore, prove that f is the Minkowski distance function for S.

30.2. Let f be a convex function defined on an open subset of E^n and let $x, y \in \mathsf{E}^n$. Prove that the directional derivative $\delta f(x; y)$ is a gauge function of y. That is,

$$\delta f(x; \lambda y) = \lambda \delta f(x; y) \qquad \text{for all } \lambda \geq 0 \text{ and } y \in \mathsf{E}^n$$

$$\delta f(x; y_1 + y_2) \leq \delta f(x; y_1) + \delta f(x; y_2) \qquad \text{for all } y_1, y_2 \in \mathsf{E}^n.$$

In particular, $\delta f(x; y)$ is a convex function of y.

30.3. (a) Let f be a convex function defined on an open subset D of E^n and let $x, y \in \mathsf{E}^n$. Prove that $\delta f(x; y) \leq f(x + y) - f(x)$.
(b) If the function f in part (a) is a gauge function (see Exercise 29.10), prove that $\delta f(x; y) \leq f(y)$.

30.4. Let f be a convex function defined on an open subset D of E^n, and let x_0, x_1, x_2, and x be points in E^n. We have seen (Exercise 30.2) that $\delta f(x_0; x)$ is a convex function of x. Now let $\delta^2 f(x_0; x_1; x_2)$ denote the directional derivative of the convex function $\delta f(x_0; x)$ at x_1 in the direction of x_2. Prove the following:
(a) $\delta^2 f(x_0; x_1; x_1) = \delta f(x_0; x_1)$.
(b) $\delta^2 f(x_0; x_1; -x_1) = -\delta f(x_0; x_1)$.
(c) x_1 is a linearity vector of $\delta^2 f(x_0; x_1; x)$. (See Exercise 29.10.)
(d) If x_2 is a linearity vector of $\delta f(x_0; x)$, then it is also a linearity vector of $\delta^2 f(x_0; x_1; x)$.

30.5. In this exercise we outline a proof of the Hahn-Banach Theorem: Let g be a semi-norm defined on E^n. (See Exercise 29.11 for the definition and basic properties of a semi-norm.) Let F be a k-dimensional subspace of E^n ($0 < k < n$). Suppose that f is a linear functional defined on F such that $|f(x)| \leq g(x)$ for all $x \in F$. Then there exists a linear functional f_1 defined on E^n such that $f_1(x) = f(x)$ if $x \in F$ and $|f_1(x)| \leq g(x)$ for all $x \in \mathsf{E}^n$.
(a) Show that the set $A \equiv \{x: g(x) < 1\}$ is an open convex set containing θ.
(b) We may assume that f is not identically zero, for otherwise we could let f_1 be identically zero and be done. Show that the set $H \equiv \{x: f(x) = 1\}$ is a $(k-1)$-dimensional flat which is disjoint from A.
(c) By Corollary 4.6, there exists a hyperplane H_0 in E^n such that $H_0 \supset H$ and $H_0 \cap A = \varnothing$. Since $\theta \in A$, $\theta \notin H_0$. Thus by Theorem 3.2, there exists a linear functional f_1 such that $H_0 = [f_1: 1]$. Show that f_1 is the desired extension of f.

*30.6. Let f be a convex function defined on an open subset D of E^n. Let A be the set of points in D at which f is not differentiable. Prove that A is a countable union of compact sets of dimension less than n.

Exercises 30.7–30.12 refer to the following **convex programming problem** P: Let U be a convex subset of E^n and suppose $f\colon U \to \mathsf{R}$ is concave. Maximize $f(x)$ on U subject to the constraints

$$h_1(x) \leq 0$$
$$\vdots$$
$$h_m(x) \leq 0$$

where each of the functions $h_i\colon U \to \mathsf{R}$ is convex.

The following terminology also applies: The **feasible set** $\mathcal{F}$ for P is defined by

$$\mathcal{F} \equiv \{x \in U : h_i(x) \leq 0 \text{ for all } i = 1,\ldots,m\}.$$

An **optimal solution** for P is a point $\bar{x} \in \mathcal{F}$ such that $f(\bar{x}) \geq f(x)$ for all $x \in \mathcal{F}$. The **Lagrangian function** $K(x, y)$ for P is defined by

$$K(x, y) \equiv f(x) - \psi_1 h_1(x) - \cdots - \psi_m h_m(x),$$

where $x \in U$ and $y \equiv (\psi_1, \ldots, \psi_m)$ is in the nonnegative orthant E^m_+ of E^m. The Lagrangian function is said to have a **saddle point** at $(\bar{x}, \bar{y}) \in U \times \mathsf{E}^m_+$ if

$$K(x, \bar{y}) \leq K(\bar{x}, \bar{y}) \leq K(\bar{x}, y)$$

for all $x \in U$ and $y \in \mathsf{E}^m_+$. A point $x \in \mathcal{F}$ is called a **strictly feasible solution** to P if $h_i(x) < 0$ for each $i = 1, \ldots, m$.

30.7. Prove that the feasible set $\mathcal{F}$ is convex.

30.8. Prove that for fixed $y \geq \theta$, $K(x, y)$ is a concave function of x on U.

30.9. Prove that for fixed $x \in U$, $K(x, y)$ is a convex function of y on E^m_+.

30.10. If $K(x, y)$ has a saddle point $(\bar{x}, \bar{y}) \in U \times \mathsf{E}^m_+$, then prove that $\bar{x}$ is an optimal solution to P.

*30.11. Suppose that P has an optimal solution $\bar{x}$ and that some solution of P is strictly feasible. Prove that there exists a $\bar{y} \in \mathsf{E}^m_+$ such that $(\bar{x}, \bar{y})$ is a saddle point for $K(x, y)$.

30.12. Define the function $g\colon \mathsf{E}^m_+ \to \mathsf{R}$ by

$$g(y) \equiv \sup_{x \in U} K(x, y) = \sup_{x \in U} \{f(x) - \psi_1 h_1(x) - \cdots - \psi_m h_m(x)\},$$

and let $\mathcal{F}^*$ be the set on which g is finite. Suppose $\mathcal{F}^* \neq \emptyset$. We define the **dual program** P^*: Minimize the convex function $g: E_+^m \to R$ on $\mathcal{F}^*$. Prove the following:
(a) g is a convex function on $\mathcal{F}^*$.
(b) If $x \in \mathcal{F}$ and $y \in \mathcal{F}^*$, then $f(x) \leq g(y)$.
(c) Suppose P has a strictly feasible solution. If P has an optimal solution $\bar{x}$, then P^* has an optimal solution $\bar{y}$ and $f(\bar{x}) = g(\bar{y})$.

SOLUTIONS, HINTS, AND REFERENCES FOR EXERCISES

SECTION 1

1.3. (b) Let $p \in \alpha(A + B)$. Then there exists $a \in A$ and $b \in B$ such that $p = \alpha(a + b) = \alpha a + \alpha b$. Thus $p \in \alpha A + \alpha B$. The converse is similar.

1.4. $A_1 + A_2 = \{(x, y): 0 \leq x \leq 2 \text{ and } 0 \leq y \leq 2\}$. $(A_1 + A_2) + A_3$ is the hexagon with vertices at θ, x_1, x_2, $(2, 4)$, $(4, 2)$, and $(4, 4)$.

1.10. (a) True. (b) False. (It is true that if A is closed and B is compact, then $A + B$ is closed.)

1.13. Let $x \in \mathsf{E}^n$ and let $\varepsilon > 0$. Then for any y such that $d(x, y) < \varepsilon$, we have

$$d(p, y) \leq d(p, x) + d(x, y) < d(p, x) + \varepsilon.$$

Thus $d(p, y) - d(p, x) < \varepsilon$. Similarly,

$$d(p, x) \leq d(p, y) + d(x, y) < d(p, y) + \varepsilon,$$

and so $d(p, x) - d(p, y) < \varepsilon$. Combining these we obtain $|d(p, x) - d(p, y)| < \varepsilon$. That is, $|f(x) - f(y)| < \varepsilon$, and it follows that f is continuous at x.

1.15. Use Theorem 1.21.

1.17. Find an open ball around x which misses A.

1.18. (a) It follows from part (b) that both sets A and B must be unbounded.

1.19. Show that if $K \neq S$, then K and $S \sim K$ are nonempty and separated.

SECTION 2

2.1.

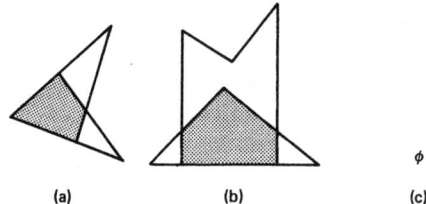

(a) (b) (c) ϕ (d)

†2.3. You might begin by looking at compact subsets of E^2.*

2.4. Let $y \in B(x, \delta)$ so that $\lambda y \in \lambda B(x, \delta)$. Then $d(x, y) < \delta$, so $d(\lambda x, \lambda y) = \lambda d(x, y) < \lambda \delta$ and $\lambda y \in B(\lambda x, \lambda \delta)$. Conversely, suppose $z \in B(\lambda x, \lambda \delta)$. Then $d(z, \lambda x) < \lambda \delta$, so

$$d\left(\frac{z}{\lambda}, x\right) = \frac{1}{\lambda} d(z, \lambda x) < \frac{1}{\lambda}(\lambda \delta) = \delta.$$

Thus $z/\lambda \in B(x, \delta)$ and $z \in \lambda B(x, \delta)$.

2.7. (a) $\{(x, y): y \geq x^2\}$.
(b) $\{(x, y): y \geq x^2 \text{ and } x > 0\} \cup \{(0, 0)\}$.
(c) E^2.
(d) $\{(x, y): 1/x \leq y < 2 \text{ and } x > \frac{1}{2}\} \cup \{(\frac{1}{2}, 2)\}$.
(e) $\{(x, y): -1 \leq y \leq 1\}$.
(f) $\{(x, y): -\pi/2 < x < \pi/2\}$.

2.10. Use Theorem 2.13.

2.13. Use Exercise 2.12 to show $\text{cl}(S) \subset \text{cl}(\text{int } S)$.

2.14. Use Exercise 2.12 to show $\text{int}(\text{cl } S) \subset \text{int } S$.

2.15. Use Exercise 2.14.

2.16. Note that $x \in (\alpha + \beta)S$ iff $\dfrac{1}{\alpha + \beta} x \in S$.

2.17. False.

2.18. Use Exercise 2.12.

2.19. Both (a) and (b) are false, since the boundary may be convex but does not have to be. For part (c) use Exercise 2.11.

†2.23. See Section 5 for results in this direction.

2.24. Pos S is a cone with vertex $(0, 0)$ containing the positive x_2 axis and with sides on the lines $x_2 = \pm x_1$.

---

*M. Breen (1981) has just proved that for any nonempty compact convex set K in E^2 there exists a compact planar set $S \neq K$ such that K is the kernel of S. Furthermore, for a nonempty closed convex set K in E^2, K is the kernel of some planar set $S \neq K$ iff K contains no line. These partial results appeared while the book was in press.

224   SOLUTIONS, HINTS, AND REFERENCES FOR EXERCISES

2.25. (a) $x = x_1 + x_2 + 2x_3 = -2x_1 + 4x_2 - x_3$. The first is a positive combination and the second is an affine combination.

2.28. (a) See the proof of Theorem 2.25.

(b) To find the affine coordinates of p with respect to $\{x_1, x_2, x_3\}$, first write $p - x_1$ as a linear combination of $x_2 - x_1$ and $x_3 - x_1$. We obtain $(\frac{12}{13}, \frac{3}{13}, \frac{-2}{13})$, $(\frac{8}{13}, \frac{2}{13}, \frac{3}{13})$, $(\frac{2}{3}, 0, \frac{1}{3})$, and $(\frac{9}{13}, \frac{-1}{13}, \frac{5}{13})$, respectively.

(c) $p_2 \in \text{int conv } T$ and $p_3 \in \text{bd conv } T$. p_1 and p_4 are not in conv T.

2.30. (a) Use Exercise 2.27(a).

(b) Show that $A \subset B$ and that A and B are flats of the same dimension. Then use Exercise 2.29.

2.31. (a) Observe that if $F \cup G$ is convex, then we can assume without loss of generality that the flats F and G are both subspaces. (Why?)

2.33. $x = \frac{7}{12}x_1 + \frac{2}{12}x_2 + \frac{3}{12}x_3$.

*2.35. See Eggleston (1966, p. 35).

*2.36. See Baker (1979).

2.37. (c) True. (d) False.

*2.38. Let $\mathbf{E}$ be the space ℓ^2 whose elements are sequences $x \equiv (x_1, x_2, \dots)$ of real numbers such that Σx_i^2 is convergent. Let S be the subset of ℓ^2 consisting of sequences that have only a finite number of nonzero terms.

*2.39. See Foland and Marr (1966).

†2.40. For results in this direction, see Breen (1978) and Stavrakas (1972).

*2.41. See McMullen (1978).

2.42. *(a) See Starr (1969), Appendix 3.

†(b) For results in this direction, see Sakuma (1977).

*2.43. This is proved by McKinney (1966). For related results see Stamey and Marr (1963), Valentine (1957), Breen (1976), and Eggleston (1975, 1976).

SECTION 3

3.2. Start with a basis for V and expand it by adding y to obtain a basis for $\mathbf{E}^n$.

3.4. (a) $u = (2, -3)$.

(b) $\alpha = -3$.

3.6. $f(x, y) = x + 4y$ and $\alpha = 7$.

3.7. x_1 and x_4 are on the same side as θ, x_3 is not. x_2 is on H.

3.8. (a) $u = (3, -1, 2, 1)$ or any multiple thereof.

(b) $f(x) = \langle u, x \rangle$ and $\alpha = 5$, where $u = (3, -1, 2, 1)$.

3.9. Either use induction on the dimension of F in decreasing order or use the fact that every flat can be considered a hyperplane in an appropriate subspace.

SOLUTIONS, HINTS, AND REFERENCES FOR EXERCISES

3.10. Use Exercises 3.9 and 2.13.
3.12. Use Corollary 3.5.
3.13. Use Exercise 2.28.

SECTION 4

4.2. Use Theorem 4.12.
4.4. Use Theorem 2.23 and 4.13.
4.6. Use Theorem 4.7.
4.7. Negative.
4.8. Use Theorem 4.7 and show that the separating hyperplane contains A.
4.9. First suppose int $S = \varnothing$. Then use Theorem 4.7 to handle the case when int $S \neq \varnothing$.
4.10. First consider the case when $A \cup B = \mathsf{E}^n$ and use Theorem 4.11.
†4.12. It is sufficient that S be open (Corollary 4.6), or compact (Theorem 4.12), or at a positive distance from F. See Klee (1951a, p. 458) for related results.

SECTION 5

5.5. Suppose x is not an extreme point of S. Then there exist $y, z \in S$ such that $x \in \text{relint}\,\overline{yz}$. Since $x \in \text{relint}\,\overline{yz}$ and H does not cut S, we must have $\overline{yz} \subset H$. Thus x is not an extreme point of $S \cap H$. Conversely, suppose x is an extreme point of S. Then x is not interior to any line segment in S and hence is not interior to any line segment in $H \cap S$. Thus x is an extreme point of $H \cap S$.
5.6. Construct a convex subset of E^3 that has exactly one line segment in its boundary.
5.9. (a) True. Use Exercise 5.5.
    (b) False. Consider the set $S = \text{conv}[\text{cl } B(\theta, 1) \cup \{(0, 2)\}]$ in E^2.
5.11. Use Exercise 5.1.
5.12. (a) Consider the profile of cl S.
5.15. (a) Let $x \in \mathsf{E}^n \sim C$ and use Theorem 4.12 to obtain a hyperplane H strictly separating x and C. Show that the hyperplane parallel to H through x_0 supports C at x_0.
    (b) Adapt the proof of Theorem 2.11.
5.18. By Exercise 2.26(a), we may assume that S is convex. To see that pos S is closed, let Ω be the surface of the unit ball in E^n and let Ψ be the central projection of $\mathsf{E}^n \sim \{\theta\}$ onto Ω. That is, for each point $p \neq \theta$ let $R(p)$ be the ray from θ through p. Then $\Psi(p) = p'$, where $R(p) \cap \Omega = \{p'\}$. Show that pos $S = \{\theta\} \cup \Psi^{-1}(\Psi(S))$. Then use the fact that Ψ is continuous to get $\Psi(S)$ compact. This implies $\Psi(S)$ is closed and so $\Psi^{-1}(\Psi(S))$ is closed in $\mathsf{E}^n \sim \{\theta\}$. It follows that $\{\theta\} \cup \Psi^{-1}(\Psi(S))$ is closed in E^n.

SECTION 6

*6.4. See Valentine (1964, p. 70).

6.7. One such point is $(1, 1) = \frac{2}{3}x_1 + \frac{1}{3}x_3 + \frac{1}{3}x_4 = \frac{1}{2}x_2 + \frac{1}{2}x_5$.

6.8. $(\frac{1}{2}, -\frac{1}{2}, 1)$

6.9. Use Helly's Theorem 6.2.

6.11. Let L_1 be a line in E^2 perpendicular to L. Project E^2 onto L_1 and apply Helly's Theorem 6.3 to the projections of $\mathcal{F}$.

6.12. There are several possible generalizations: One may require pairs of sets to intersect and replace L by a hyperplane. One may require every n sets to intersect and leave L as a line. More generally, one may require every k sets to intersect $(2 \leq k \leq n)$ and replace L by an $(n - k + 1)$-dimensional flat.

6.13. Show that the existence of a point in $\mathrm{conv}\, S_1 \cap \mathrm{conv}\, S_2$ would contradict the affine independence of S.

*6.14. See Horn (1949) or Valentine (1964, p. 73).

*6.15. See Danzer, Grünbaum, and Klee (1963) for a good discussion of Helly's Theorem and how it relates to other important results.

*6.16. See Borwein (1977).

*6.17. See Valentine (1970).

*6.18. See Buchman and Valentine (1976).

*6.19. See Breen (1979).

6.20. (a) Without loss of generality, we may assume the line segments are vertical. That is, $\mathcal{F} \equiv \{A_i: i = 1, \ldots, m\}$, where $A_i \equiv \{(x, y): x = x_i, u_i \leq y \leq v_i\}$. Consider the sets $B_i \equiv \{(m, b): u_i \leq mx_i + b \leq v_i\}$.

*(b) See Grünbaum (1958).

*(c) See Katchalski and Lewis (1980).

†(d) Katchalski and Lewis (1980) have conjectured that $k = 2$ is the smallest. See Danzer, Grünbaum, and Klee (1963) for related results.

†(e) Grünbaum (1958) has conjectured that $m = 5$ is the smallest such number.

SECTION 7

7.2. (a) $\delta_{123} = \frac{3}{4}, \delta_{124} = \frac{2}{3}, \delta_{134} = \frac{1}{6}, \delta_{234} = \frac{1}{2}$.
(b) $y = \frac{1}{2}x + \frac{5}{4}$.

7.3. Let $S \equiv \{x: \|x\| \leq \delta/2\}$ and show that the two sets $P + S$ and $Q + S$ can be strictly separated by a hyperplane.

*7.4. Use Theorem 4.11 and Caratheodory's Theorem 2.23.

*7.8. See Watson (1973).

SOLUTIONS, HINTS, AND REFERENCES FOR EXERCISES

SECTION 8

- **8.1.** Use Caratheodory's Theorem 2.23.
- **8.3.** Use Exercise 7.4 and the projection idea in the proof of Theorem 8.2.
- ***8.5.** See Valentine (1964, p. 41).

SECTION 9

- **9.2.** Note that when P consists of a single point and $k = n$, then the $(n-1)$-cylinder generated by P is precisely a hyperplane.
- **9.3.** Use the following result from linear algebra: If S and T are subspaces, then $\dim S + \dim T = \dim(S \cap T) + \dim(S + T)$.
- ***9.4.** Use Exercise 9.3.
- ***9.5.** See Lay (1971a, p. 31).
- ***9.8.** See Molnár (1957).
- ***9.9.** The proof uses Exercise 9.8. See Baker (1967a). For related (more general) results, see Baker (1967b) and Katchalski (1977).

SECTION 10

- **10.1.** Project onto a line orthogonal to H and use Helly's Theorem 6.3.
- **†10.4.** At this time (1982) there is nothing in the literature beyond what is included in this chapter. You're on your own!

SECTION 11

- **11.2.** $120°$
- **11.3.** $\frac{1}{2}(\pi - \sqrt{3})w^2$
- **11.6.** Suppose there are two circumcircles and use their intersection to obtain another of smaller radius.
- **11.10.** Use Corollary 11.4.
- ***11.11.** See Yaglom and Boltyanskii (1961, pp. 250–253).
- ***11.12.** See Eggleston (1966, p. 124).
- ***11.13.** See Besicovitch (1963).
- ***11.14.** See Danzer (1963).
- ***11.15.** See Klamkin (1979) and Fejes-Toth (1964, p. 312).

SECTION 12

- **12.1.** (a) Use Theorem 11.3.
- **12.2.** Use Theorem 6.5.

*12.5. See Eggleston (1966, p. 126).

12.7. Use Exercise 12.5 to assume S has constant width 1. Consider the points at which the incircle intersects the boundary and connect three of these (properly chosen) to the center of the incircle.

12.8. Use Theorem 12.4 to cover S by a regular hexagon of minimum width d. Then divide this hexagon into three congruent pentagons.

†12.9. For a discussion of this problem, see Eggleston (1955, pp. 11–24; 1957, pp. 77–92).

*12.10. See Grünbaum (1963, p. 274) and Pal (1920).

SECTION 13

13.2. Use Exercise 13.1.

13.3. Use Theorem 13.4.

*13.5. See Yaglom and Boltyanskii (1961, p. 209).

†13.6. This appears as Problem 23 in Moser (1980). For k odd, it is known (Reinhardt, 1922) that the unique maximizing k-gon is regular. For k even, the problem is not so simple. For $k = 4$, the maximum area of $\frac{1}{2}$ is attained for infinitely many quadrilaterals, each with perpendicular diagonals of length 1. See Graham (1975) for $k = 6$. Nothing is known for even $k > 6$.

†13.7. This appears as Problem 6 in Moser (1980). It is known that $\frac{1}{3} \leq f(3) \leq \sqrt{2}/4$.

†13.8. For a general discussion of the isoperimetric inequality, including a partial answer to this exercise, see Osserman (1978).

†13.9. This is a well-known unsolved problem. See Valentine (1964, p. 192) for a discussion of incorrect proofs and partial results.

*13.10. See Süss (1926).

*13.11. See Kelly (1963).

*13.12. See Chakerian and Lange (1971).

SECTION 14

14.2. $D(A, B) = 3$, $D(A, C) = 3\sqrt{2} - 1$, $D(B, C) = 3$.

14.4. $D(\{p\}, S) = d(x, p) + \delta$.

14.5. $D(S_1, S_2) = d(x, y) + |\alpha - \beta|$.

SECTION 15

15.1. Use Theorem 1.21 and look at the sets $E^n \sim B_k$.

15.2. Use the Bolzano-Weierstrass theorem from analysis.

*15.3. See Beer (1975).

SECTION 16

16.1. Let $r(F)$ and $c(F)$ be the radius and center of F. Then show (as in Theorem 16.1) that r and c are continuous functions on $\{F_i\}$. Let $r' = \lim r(F_i)$ and $c' = \lim c(F_i)$. Then show that G must be the sphere of radius r' centered at c'.

*16.4. See Eggleston (1966, p. 111).

†16.5. Valentine (1964, p. 187) has conjectured that the maximal set will be a region cut off from S by two parallel planes equidistant from the center.

SECTION 17

17.3. This exercise comes from Valentine (1964, p. 93).

*17.4. See Valentine (1946 or 1964, p. 91).

SECTION 18

18.3. $x_1, x_2, x_3,$ and x_4.

SECTION 19

19.5. See McMullen and Shephard (1971, pp. 31–38).

†19.6. See Klee (1961) for a discussion.

SECTION 20

20.1. (b) Let $S \equiv \text{conv}[\text{cl } B(\theta, 1) \cup \{(3, 1)\}]$. Then $(0, 1)$ is an extreme point of S but it is not a vertex of S.

20.2. Use Theorems 20.8 and 20.9.

20.5. (c) Begin with the case $k = 2$. That is, let $P = P_1 + P_2$. Show that each vertex of P is the algebraic sum of a vertex of P_1 and a vertex of P_2.

20.6. Use Theorems 20.8 and 20.9 and look at the corresponding closed half-spaces.

20.8. (a) $f_{124}(x, y, z) = 3y - z$, $f_{134}(x, y, z) = 3x - z$, $f_{123}(x, y, z) = z$.

20.9. (b) Two polygons in E^2 are combinatorially equivalent iff they have the same number of vertices.

†(c) See Grünbaum (1967), Chapter 13, for related results when $n = 3$.

SECTION 21

21.1. Each facet of X^4 is the convex hull of a facet of X^3 and one of the apex points. Use this fact to list them.

21.3. Note that $\binom{a}{b+1} + \binom{a}{b} = \binom{a+1}{b+1}$.

21.7. Suppose that a regular polyhedron has r facets each of which is a k-sided regular polygon and that s edges meet at each vertex. Letting v and e denote the number of vertices and edges in the polyhedron, show that $kr = 2e$ and $sv = 2e$. Since $v - e + r = 2$, obtain $2e/s + 2e/k - e = 2$, so that $1/s + 1/k = 1/2 + 1/e$. Find all the integral solutions of this last equation that satisfy the geometric constraints of the problem.

*21.8. This is proved in Coxeter (1963, p. 136). If we call a regular polytope with m facets an m-cell, then in E^4 we have the 5-cell with tetrahedral facets (the simplex), the 8-cell with cubic facets (the hypercube), the 16-cell with tetrahedral facets (the regular cross-polytope), the 24-cell with octahedral facets, the 120-cell with dodecahedral facets, and the 600-cell with tetrahedral facets. In dimensions greater than four, only the simplex, the hypercube, and the regular cross-polytope are regular. See Hilbert and Cohn-vossen (1952, pp. 143-157) for a related discussion.

SECTION 22

22.1. $\delta = \sec(\pi/k)$.

22.3. $k_2 = 4$ and $k_4 = 11$.

22.4. The case for $n = 2$ is well-known and may be assumed. For higher dimensions, use induction on n.

*22.5. See Benson (1966, p. 188).

*22.8. See Benson (1966, p. 12).

22.10. †(a) See Sallee (1969).
†(b) See Klee (1979).

SECTION 23

23.1. (a) The half-plane $\{(x, y): x \leq \frac{1}{2}\}$.
(b) The half-plane $\{(x, y): x \leq 0\}$.

23.2. Use Caratheodory's Theorem 2.23.

23.4. See Theorem 23.3(b) and the proof of Theorem 23.5.

*23.11. Consider an n-polytope P^* dual to P. Then each k-face of P corresponds to an $(n - 1 - k)$-face of P^*. Apply Euler's formula to the face F^*. See Grünbaum (1967, p. 137).

SECTION 24

24.1. Suppose $x \notin M$ and reach a contradiction.

24.3. Use Theorem 4.12 and Theorem 24.6.

SOLUTIONS, HINTS, AND REFERENCES FOR EXERCISES

*24.4. See Valentine (1963, p. 488 or 1964, p. 88).
*24.6. See Valentine (1963, p. 482) for a proof using duality.
*24.7. See Valentine (1963, p. 479 or 1964, p. 73).

SECTION 25

25.1. (a) $E(x, y) = -\frac{1}{4}$, $v(x) = -\frac{4}{3}$, $v(y) = \frac{1}{4}$.
25.3. (a) $\bar{x} = (0, 1)$, $\bar{y} = (1, 0)$, $v = 4$.
(c) $\bar{x} = (0, 1)$, $v = 2$, and any strategy is optimal for P_{II}.
(h) $\bar{x} = (0, 0, \frac{5}{8}, \frac{3}{8})$, $\bar{y} = (0, \frac{3}{8}, \frac{5}{8}, 0)$, $v = \frac{31}{8}$.
25.5. (a) $\bar{x} = \left(\dfrac{d-c}{d-c+a-b}, \dfrac{a-b}{d-c+a-b} \right)$.
25.7. (a) $\bar{x} = (\frac{1}{3}, \frac{2}{3})$. (b) $\bar{x} = (\frac{4}{7}, \frac{3}{7})$.
25.11. Assume the value of the game $v = 0$ and use Exercise 25.9 and Theorem 20.9.
*25.13. See Gale (1960, p. 225).

SECTION 26

26.1. (a) $x_1 = \frac{72}{5}$, $x_2 = \frac{16}{5}$, max $= 1376$.
(c) unbounded.
26.3. (d) Minimize $\quad 6y_1 - 4y_2$
subject to $\quad 2y_1 - y_2 \geq 2$
$\quad\quad\quad\quad\quad -2y_1 + 2y_2 \geq 5$
and $y_1 \geq 0$, $y_2 \geq 0$.
This program is unbounded.
26.5. Prove the contrapositive.
26.7. See Proof 26.12.
26.8. (a) 20 widgets and 30 whammies. (b) Below $.80.

SECTION 27

27.1. (a) $x_1 = 12$, $x_2 = 2$, max $= 44$.
(b) $x_1 = 7$, $x_2 = \frac{23}{2}$, $x_3 = \frac{1}{2}$, max $= \frac{147}{2}$.
(c) $x_1 = 10$, $x_4 = 5$, $x_2 = x_3 = x_5 = 0$, max $= 350$.
(d) $x_1 = 0$, $x_2 = \frac{9}{2}$, $x_3 = 6$, min $= \frac{21}{2}$.
27.4. (x_1, x_2, x_3) will be an optimal solution iff it is a convex combination of $(8, 4, 0)$ and $(0, 12, 8)$.
27.5. Solve the dual to the given problem to obtain 11 bushels of oats and 3 bushels of barley at a cost of $6.70.
27.6. 10 bottles of Musty, 100 bottles of Crusty, and no Dusty or Rusty for a maximum profit of $680.

SECTION 28

27.7. $6000 in mutual funds, $4000 in bonds, $2000 in savings for a return of $1100.

27.8. $6000 in mutual funds and $6000 in savings for a return of $1020.

27.9. Use Proof 26.12 to change this into a linear programming problem and then use the simplex method. He should invest $11,000 in stocks, $9000 in bonds, and $15,000 in gold.

SECTION 28

28.2. Use Theorem 1.4.

28.4. Let the sides have length w, x, y, z, and consider $f[(w + x + y + z)/4]$, where $f(t) = t^2$.

28.6. (a) True. (b) False.

28.8. Evaluate the convex function $f(t) = e^{2t}$ at the point $\frac{1}{2}(x + y)$.

28.9. Use Jensen's Inequality (Theorem 28.6) and the convex function $f(t) = -\ln t$.

28.11. To show $g(\alpha x + \beta y) \leq \alpha g(x) + \beta g(y)$, first consider the case when either x or y is θ. Then suppose $x \neq \theta$ and $y \neq \theta$. Show that

$$f\left(\frac{x}{g(x)}\right) = f\left(\frac{y}{g(y)}\right) = 1,$$

so that

$$f\left(\frac{x+y}{g(x)+g(y)}\right) = f\left(\frac{g(x)}{g(x)+g(y)}\frac{x}{g(x)} + \frac{g(y)}{g(x)+g(y)}\frac{y}{g(y)}\right)$$

$$\leq 1.$$

Then look at

$$g\left(\frac{x+y}{g(x)+g(y)}\right)$$

and use the fact that $g(\lambda x) = \lambda g(x)$ for $\lambda \geq 0$.

28.12. Let $x \equiv (\alpha_1, \ldots, \alpha_m)$ and show that the function defined by $f(x) \equiv \sum_{i=1}^{m} \alpha_i^p$ is convex and homogeneous of degree p on the nonnegative orthant in $\mathbf{E}^m$. Then use Exercise 28.11.

*28.17. (a) See Roberts and Varberg (1973, p. 212).
(b) See Valentine (1964, p. 131).

SECTION 29

29.2. First consider the case where $n = 2$ and draw a picture. How many extreme points does A have?

SOLUTIONS, HINTS, AND REFERENCES FOR EXERCISES

29.7. $A \subset B$ follows easily from the definitions. For the converse, suppose there exists $p \in A \sim B$ and obtain a hyperplane strictly separating $\{p\}$ and B.

29.9. Suppose $q(x)$ is subadditive and positively homogeneous. Show that $S = \{x: q(x) \leq 1\}$ and use this to show that S is convex.

SECTION 30

30.1. Show that f is convex and that S is convex. Then use Theorem 30.2 to show that S is closed and that $\theta \in \text{int } S$. If S were not bounded, then S would contain a ray from θ. Show that this leads to a contradiction.

30.2. For subadditivity, show $f(x + \alpha(y_1 + y_2)) \leq \frac{1}{2}f(x + 2\alpha y_1) + \frac{1}{2}f(x + 2\alpha y_2)$. Then subtract $f(x)$, divide by α, and take limits.

30.4. (a) Use the fact that $\delta f(x_0; x)$ is a gauge (Exercise 30.2).

30.5. (a) Use Theorem 30.2 to show that A is open.
(c) Show that if $|f_1(y)| > g(y)$ for some $y \in E^n$, then $y/f_1(y) \in A \cap H_0$.

*30.6. See Stoer and Witzgall (1970, p. 153) or Anderson and Klee (1952).

*30.11. See Roberts and Varberg (1973, p. 172).

BIBLIOGRAPHY

Anderson, R., and V. Klee. 1952. Convex functions and upper semi-continuous collections, *Duke Math. J.*, **19**, 349–357.

Baker, M. J. C. 1967. A Helly-type theorem on a sphere, *J. Austral. Math. Soc.*, **7**, 323–326. (a)

———. 1967. A spherical Helly-type theorem, *Pacific J. Math.*, **23**, 1–3. (b)

———. 1979. Covering a polygon with triangles: A Caratheodory-type theorem, *J. Austral. Math. Soc. (Series A)*, **28**, 229–234.

Barbier, E. 1860. Note sur le problème de l'aiguille et le jeu joint couvert, *J. Math. Pures Appl.*, ser. 2, **5**, 273–286.

Beer, G. 1975. Starshaped sets and the Hausdorff metric, *Pacific J. Math.*, **61**(1), 21–27.

Benson, R. V. 1966. *Euclidean Geometry and Convexity*, McGraw-Hill, New York.

Besicovitch, A. S. 1963. On semicircles inscribed into sets of constant width, Am. Math. Soc. Symposium on Convexity, *Proc. Symp. Pure Math.*, **7**, 15–18.

Bonnesen, T., and W. Fenchel. 1934. *Theorie der konvexen Körper*, Springer-Verlag, Berlin; rp. Chelsea, New York, 1948.

Borsuk, K. 1933. Drei Sätze über die n-dimensionale euklidische Sphäre, *Fund. Math.*, **20**, 177–190.

Borwein, J. 1977. A proof of the equivalence of Helly's and Krasnoselski's theorems, *Canad. Math. Bull.*, **20**(1), 35–37.

Botts, T. 1942. Convex sets, *Am. Math. Monthly*, **49**, 527–535.

Breen, M. 1976. A decomposition theorem for m-convex sets, *Israel J. Math.*, **24**(3–4), 211–216.

———. 1978. Sets in R^d having $(d-2)$-dimensional kernels, *Pacific J. Math.*, **75**(1), 37–44.

———. 1979. A Helly-type theorem for the dimension of the kernel of a starshaped set, *Proc. Am. Math. Soc.*, **73**(2), 233–236.

———. 1981. Admissible kernels for starshaped sets, *Proc. Am. Math. Soc.*, **82**(4), 622–628.

Buchman, E., and F. A. Valentine. 1976. External visibility, *Pacific J. Math.*, **64**(2), 333–340.

Caratheodory, C. 1907. Über den Variabilitätsbereich der Koeffizienten von Potenzreihen, die gegebene Werts nicht annahmen, *Math. Ann.*, **64**, 95–115.

BIBLIOGRAPHY

Chakerian, G. D., and L. H. Lange. 1971. Geometric extremum problems, *Math. Mag.*, **44**, 57-69.

Charnes, A. 1952. Optimality and degeneracy in linear programming, *Econometrica*, **20**, 160-170.

Coxeter, H. S. M. 1963. *Regular Polytopes*, 2nd ed., Macmillan, New York.

Danzer, L. W. 1963. A characterization of the circle, Am. Math. Soc. Symp. on Convexity, *Proc. Symp. Pure Math.*, **7**, 99-100.

Danzer, L. W., G. Grünbaum, and V. Klee. 1963. Helly's theorem and its relatives, Am. Math. Soc. Symp. on Convexity, *Proc. Symp. Pure Math.*, **7**, 101-180.

DeSantis, R. 1957. A generalization of Helly's theorem, *Proc. Am. Math. Soc.*, **8**, 336-340.

Dieudonné, J. 1941. Sur Théorème de Hahn-Banach, *Rev. Sci.*, **79**, 642-643.

Dines, L. L., and N. H. McCoy. 1933. On linear inequalities, *Trans. Roy. Soc. Canada*, Sec. III, **27**, 37-70.

Eggleston, H. G. 1955. Covering a three-dimensional set with sets of smaller diameter, *J. London Math. Soc.*, **33**, 11-24.

_____. 1957. *Problems in Euclidean Space: Applications of Convexity*, Pergamon, New York.

_____. 1966. *Convexity*, Cambridge University Press, Cambridge, England.

_____. 1975. A proof of a theorem of Valentine, *Math. Proc. Cambridge Philos. Soc.*, **77**, 525-528.

_____. 1976. Valentine convexity in n dimensions, *Math. Proc. Cambridge Philos. Soc.*, **80**, 223-228.

Fejes-Tóth, L. 1964. *Regular Figures*, Pergamon, London.

Foland, N. E., and J. M. Marr. 1966. Sets with zero-dimensional kernels, *Pacific J. Math.*, **19**(3), 429-432.

Gale, D. 1960. *The Theory of Linear Economic Models*, McGraw-Hill, New York.

Gelbaum, B., and J. Olmstead. 1964. *Counterexamples in Analysis*, Holden-Day, San Francisco.

Golstein, E. G. 1972. *Theory of Convex Programming*, American Math. Soc., Providence, Rhode Island.

Graham, R. L. 1975. The largest small hexagon, *J. Comb. Theory A*, **18**, 165-170.

Grünbaum, B. 1958. On common transversals, *Arch. Math.*, **9**, 465-469.

_____. 1963. Borsuk's problem and related questions, Am. Math. Soc. Symp. on Convexity, *Proc. Symp. Pure Math.*, **7**, 271-284.

_____. 1967. *Convex Polytopes*, Wiley, New York.

Hadwiger, H., H. Debrunner, and V. Klee. 1964. *Combinatorial Geometry in the Plane*, Holt, New York.

Hanner, O., and H. Rådström. 1951. A generalization of a theorem of Fenchel, *Proc. Am. Math. Soc.*, **2**, 589-593.

Helly, E. 1923. Über Mengen konvexer Körper mit gemeinschaftlichen Punkten, *Jahrb. Deut. Math. Verein*, **32**, 175-176.

Hilbert, D., and S. Cohn-vossen. 1952. *Geometry and the Imagination*, Chelsea, New York. (English translation of their 1932 book *Anschauliche Geometrie*.)

Hille, E., and R. Phillips. 1957. Functional Analysis and Semigroups, *Am. Math. Soc. Colloq. Publ.*, No. 31.

Horn, A. 1949. Some generalizations of Helly's theorem on convex sets, *Bull. Am. Math. Soc.*, **55**, 923–929.

Katchalski, M. 1977. A Helly-type theorem on the sphere, *Proc. Am. Math. Soc.*, **66**(1), 119–122.

Katchalski, M., and T. Lewis. 1980. Cutting families of convex sets, *Proc. Am. Math. Soc.*, **79**(3), 457–461.

Kelly, P. J. 1963. Plane convex figures, *Enrichment Math. for High School, Twenty-eighth Yearbook*, National Council of Teachers of Mathematics, Washington, D. C., 251–264.

Kelly, P. J., and M. L. Weiss. 1979. *Geometry and Convexity*, Wiley-Interscience, New York.

Kirchberger, P. 1903. Über Tschebyschefsche Annäherungsmethoden, *Math. Ann.*, **57**, 509–540.

Klamkin, M. S. 1979. The circumradius-inradius inequality for a simplex, *Math. Mag.*, **52**(1), 20–22.

Klee, V. L. 1951. Convex sets in linear spaces, *Duke Math. J.*, **18**(2), 443–466. (a)

———. 1951. On certain intersection properties of convex sets, *Canad. J. Math.*, **3**, 272–275. (b)

———. 1953. The critical set of a convex body, *Amer. J. Math.*, **75**, 178–188.

———. 1961. Convexity on Chebychev sets, *Math. Ann.*, **142**, 292–304.

———. 1979. Some unsolved problems in plane geometry, *Math. Mag.*, **52**(3), 131–145.

Krasnosselsky, M. A. 1946. Sur un Critère pour qu'un domain soit étoilé (Russian with French summary), *Mat. Sb.*, N.S., **19**(61), 309–310.

Lay, S. R. 1971. *Combinatorial Separation Theorems and Convexity* (Dissertation), University of California at Los Angeles. (a)

———. 1971. On separation by spherical surfaces, *Am. Math. Monthly*, **78**(10), 1112–1113. (b)

———. 1972. Separation by cylindrical surfaces, *Proc. Am. Math. Soc.*, **36**(1), 224–228.

———. 1980. Separating two compact sets by a parallelotope, *Proc. Am. Math. Soc.*, **79**(2), 279–284.

Lyusternik, L. A. 1963. *Convex Figures and Polyhedra*, Dover, New York.

Marcus, M., and H. Mink. 1966. *Modern University Algebra*, Macmillan, New York.

McKinney, R. L. 1966. On unions of two convex sets, *Can. J. Math.*, **18**, 883–886.

McMullen, P. 1978. Sets homothetic to intersections of their translates, *Mathematika*, **25**, 264–269.

McMullen, P., and G. C. Shepard. 1971. *Convex Polytopes and the Upper Bound Conjecture*, Cambridge University Press, Cambridge, England.

Minkowski, H. 1911. Theorie der konvexen Körper, insbesondere Begründung ihres Oberflächenbegriffs, *Ges. Abhandl., Leipzig-Berlin*, **2**, 131–229.

Molnár, J. 1957. Über eine Übertragung des Hellyschen Satzes in sphärische Räume, *Acta Math. Acad. Sci. Hungary*, **8**, 315–318.

Moser, W. 1980. *Problems in Discrete Geometry*, 5th ed., Department of Mathematics, McGill University, Montreal, Quebec.

Motzkin, T. S. 1935. Sur quelques propriétés caractéristiques des ensembles convexes, *Rend. Reale Acad. Lincei, Classe Sci. Fis., Math. Nat.*, **21**, 773–779.

Motzkin, T. S., E. G. Straus, and F. A. Valentine. 1953. The number of farthest points, *Pacific J. Math.*, **3**, 221–232.

Nakajima, S. (same as S. Matsumura). 1928. Über konvexe Kurven und Flächen, *Tohoku Math. J.*, **29**, 227–320.

Osserman, R. 1978. The isoperimetric inequality, *Bull. Am. Math. Soc.*, **84**(6), 1182–1238.

Pál, J. 1920. Über ein elementares Variationsproblem, *Danske Vid. Selsk. Math.-Fys. Medd*, **3**(2).

Prenowitz, W., and J. Jantosciak. 1979. *Join Geometries: A Theory of Convex Sets and Linear Geometry*, Springer-Verlag, New York.

Rademacher, H., and I. J. Schoenberg. 1950. Helly's theorem on convex domains and Tchebycheff's approximation problem, *Can. J. Math.*, **2**, 245–256.

Radon, J. 1921. Mengen konvexer Körper, die einen gemeinsamen Punkt enthalten, *Math. Ann.*, **83**, 113–115.

Reinhardt, K. 1922. Extremale Polygone mit gegebenen Durchmessers, *Jahr. dtsch. Math.* **31**, 251–270.

Reuleaux, F. 1875. *Lehrbuch der Kinematik*, Brunswick (Braunschweig, Germany).

Roberts, A. W., and D. E. Varberg. 1973. *Convex Functions*, Academic, New York.

Rockafellar, R. T. 1970. *Convex Analysis*, Princeton University Press, Princeton, New Jersey.

Rudin, W. 1976. *Principles of Mathematical Analysis*, 3rd ed., McGraw-Hill, New York.

Sakuma, Itsuo. 1977. Closedness of convex hulls, *J. Economic Theory*, **14**, 223–227.

Sallee, G. T. 1969. Are equidecomposable plane convex sets convex equidecomposable?, *Am. Math. Monthly*, **76**, 926–927.

Sandgren, L. 1954. On convex cones, *Math. Scand.*, **2**, 19–28; correction, **3**, page 170 (1955).

Shimrat, M. 1955. Simple proof of a theorem of P. Kirchberger, *Pacific J. Math.*, **5**, 361–362.

Stamey, W. L., and J. M. Marr. 1963. Union of two convex sets, *Can. J. Math.*, **15**, 152–156.

Starr, R. M. 1969. Quasi-equilibria in markets with non-convex preferences, *Econometrica*, **37**, 25–38.

Stavrakas, N. M. 1972. The dimension of the convex kernel and points of local nonconvexity, *Proc. Am. Math. Soc.*, **34**(1), 222–224.

Steiner, J. 1938. Einfache Beweise der isoperimetrischen Hauptsätze, *J. Reine Angew. Math.*, **18**, 289–296.

Stoer, J., and C. Witzgall. 1970. *Convexity and Optimization in Finite Dimensions I*, Springer-Verlag, New York.

Süss, W. 1926. Eine charakteristische Eigenschaft der Kugel, *Jahrb. Deut. Math. Verein*, **34**, 245–247.

Tietze, H. 1928. Über Konvexheit im kleinen und im grossen und über gewisse den Punkten einer Menge zugeordnete Dimensionszahlen, *Math. Z.*, **28**, 697–707.

_____. 1929. Bemerkungen über konvexe und nicht-konvexe Figuren, *J. Reine Angew. Math.*, **160**, 67–69.

Valentine, F. A. 1946. Set properties determined by conditions on linear sections, *Bull. Am. Math. Soc.*, **10**, 100–108.

———. 1957. A three-point convexity property, *Pacific J. Math.*, **7**, 1227–1235.

———. 1960. Characterizations of convex sets by local support properties, *Proc. Am. Math. Soc.*, **11**, 112–116.

———. 1963. The dual cone and Helly type theorems, Am. Math. Soc. Symp. on Convexity, *Proc. Symp. Pure Math.* **7**, 473–493.

———. 1964. *Convex Sets*, McGraw-Hill, New York.

———. 1970. Visible shorelines, *Am. Math. Monthly*, **77**(2), 146–152.

von Neumann, J., and O. Morgenstern. 1944. *Theory of Games and Economic Behavior*, Princeton University Press, Princeton, New Jersey.

Watson, D. 1973. A refinement of theorems of Kirchberger and Caratheodory, *J. Austral. Math. Soc.*, **15**, 190–192.

Yaglom, I. M., and V. G. Boltyanskii. 1961. *Convex Figures*, Holt, New York.

AUTHOR INDEX

Anderson, R., 233

Baker, M. J. C., 224, 227
Barbier, E., 81
Beer, G., 228
Benson, R. V., 230
Besicovitch, A. S., 227
Blaschke, 82, 98
Boltyanskii, V. G., 81, 227, 228
Borsuk, K., 88
Borwein, J., 226
Breen, M., 223, 224, 226
Buchman, E., 226

Caratheodory, C., 17, 25, 56
Chakerian, G, D., 228
Charnes, A., 190
Cohn-Vossen, S., 230
Coxeter, H. S. M., 116, 130, 230

Dantzig, G., 154
Danzer, L. W., 226, 227
Dieudonné, J., 146
Dines, L. L., 63

Eggleston, H. G., 51, 224, 227, 228, 229
Euler, L., 78, 123, 131

Fejes-Toth, L., 227
Foland, N. E., 224

Gale, D., 88, 231
Gelbaum, B., 216
Goldstein, E. G., 198
Graham, R. L., 228
Grünbaum, B., 116, 122, 130, 226, 228, 229, 230

Hanner, O., 69
Helly, 47-50, 56
Hilbert, D., 230

Hille, E., 146
Horn, A., 51, 52, 226

Katchalski, M., 226, 227
Kelly, P. J., 228
Kirchberger, P., 56
Klamkin, M. S., 227
Klee, V, L., 50, 104, 146, 225, 226, 229, 230, 233
Krasnosselsky, M. A., 52, 53

Lange, L, H., 228
Lay, S. R., 227
Lewis, T., 226

Marr, J. M., 224
McCoy, N. H., 63
McKinney, R. L., 224
McMullen, P., 116, 224, 229
Molnár, J., 227
Morgenstern, 160
Moser, W., 228
Motzkin, T. S., 112

Nakajima, S., 104

Olmstead, J., 216
Osserman, R., 228

Pál, J., 228
Phillips, R., 146

Rademacher, H., 56
Radon, J., 47
Radström, H., 69
Reinhardt, K., 228
Reuleaux, F., 78, 81
Roberts, A. W., 198, 218, 232, 233

239

AUTHOR INDEX

Rockafellar, R. T., 198, 218
Rudin, W., 8

Sakuma, I., 224
Salle, G. T., 230
Sandgren, L., 146
Schoenberg, I. J., 56
Shepherd, G. C., 116, 229
Shimrat, M., 56
Stamey, W. L., 224
Starr, R. M., 224
Stavrakas, N. M., 224
Steiner, J., 89

Stoer, J., 198, 233

Tietze, H., 104, 110

Valentine, F. A., 61, 107, 134, 146, 147, 224, 226, 227, 228, 229, 231, 232
Varberg, D. E., 198, 218, 232, 233
von Neumann, J., 160

Watson, D., 226
Witzgall, C., 198, 233

Yaglom, I. M., 81, 227, 228

SUBJECT INDEX

Affine:
  combination, 14
  coordinates, 24
  dependence, 15, 16, 47
  hull, 16, 17, 23
  independence, 15
  set, 14, 43
Apex of pyramid, 125
Area, 81, 102, 133

Ball, 4
Barbier's theorem, 81
Barycentric coordinates, 19
Basic solution, 184
Basic variable, 185
Basis:
  of bypyramid, 127
  of prism, 129
  of pyramid, 125
  of subspace, 12, 43
β-box, 70
Binomial coefficient, 125
Bipolar theorem, 146
Bipyramid, 127, 143
Blaschke selection theorem, 98
Body, 133
Bound, 37
Boundary, 8, 23
Bounded set, 5, 22

Canonical linear programming problem, 168
Caratheodory's theorem, 17, 20, 25, 74
Cauchy sequence, 98
Center, 92
Centrally symmetric set, 213
Centroid, 19
Chord, 90
Circumcircle, 79
Circumsphere, 103
Closed set, 4, 16, 21

Closure of set, 5, 13
Combinatorily equivalent, 123
Compact set, 8, 21
Compact family, 100
Complement, 4
Component, 8
Concave function, 198
Cone, 45, 149, 206
Continuous function, 6, 100, 102, 136, 215
Convergence, 96, 98, 100
Convex:
  body, 133
  combination, 14, 15
  equidecomposable, 139
  function, 198, 203
  hull, 16, 17, 21, 23, 43
  programming problem, 220
  set, 11, 15, 42, 96, 106, 109, 110, 112
Corner, 78
Crosscut, 107
Crosspolytope, 126
Cut, 37
Cycling, 190, 193
Cylinder, 64

Decomposition, 138
Deviation, 58
Diameter, 76
Dido's problem, 91
Differentiable function, 218
Dimension, 12
Dircetional derivative, 217
Distance, 3, 4, 10, 95
Distance function, 210
Dominate, 163
Dual:
  cone, 147, 213
  polytope, 142, 145
  programming problem, 173, 221
Duality, 140

241

SUBJECT INDEX

Duality theorem, 174

Edge, 116
Epigraph, 199, 213
Equichordal point, 92
Equidecomposable, 138
Equilibrium point, 155
Equilibrium theorem, 179
Equivalent by finite decomposition, 138
Euclidean space, 1
 characteristic, 130
 formula, 123, 130, 131, 132, 146
Expected payoff, 157
Exposed point, 45, 108, 116
Extreme point, 42, 44, 45, 117, 121, 171

Face, 116, 118, 121
Facet, 116
Farkas lemma, 174
Feasible set, 169, 220
Feasible solution, 169, 184
Flat, 12, 14, 16
Fréchet differentiable, 218
Function:
 concave, 198
 continuous, 6, 100, 102, 136, 215
 convex, 198, 203
 differentiable, 218
 distance, 210
 gauge, 214
 homogeneous, 205
 Lagrangian, 220
 linear, 27, 58
 mid-convex, 205
 Minkowski, 210
 objective, 168
 positively homogeneous, 206
 quasiconvex, 205
 subadditive, 212
 support, 206
Fundamental theorem for matrix games, 160 181

Gateaux differentiable, 218
Gauge function, 214

Hahn-Banach theroem, 219
Half-line, 45
Half-space, 37
Hausdorff metric, 96
Heine-Borel theorem, 8
Helly's theorem, 48, 50, 152
Homogeneous function, 205
Homothetic, 7, 22

Horn's theorem, 52
Hull:
 affine, 16, 17, 23
 convex, 16, 17, 21, 23, 43
 positive, 23, 46
Hypercube, 128
Hyperplane, 12, 27, 31
Hyperplane of support, 41, 45, 77, 116

Improper face, 116
Improper separation, 38
Incircle, 79, 101
Infeasible program, 170
Inner product, 1, 2
Interior of a set, 5, 13
Interior point, 4
Isomorphic polytopes, 123
Isoperimetric inequality, 91
Isoperimetric problem, 88

Jensen's inequality, 203

K-Cylinder, 64
Kernel, 11, 22, 25, 55
K-face, 116
Kirchberger's theorem, 56, 60, 70, 152
K-point simplicial property, 56, 59, 69
K-polytope, 116
Krasnosselsky's theorem, 53

Lagrangian function, 220
Lebesgue covering problem, 87
Limit of a sequence, 96
Line, 12
Linear combination, 12
Linear function, 27, 58
Linear functional, 27, 28, 44
Linearity vector, 214
Linearly dependent, 12, 16
Linearly independent, 12, 43
Linear programming, 168
Linear space, 1
Line segment, 10
Local concavity, 108
Locally convex set, 105
Local support, 108

Matrix game, 157
Metric, 96
Metric space, 4, 100, 115
Mid-convex function, 205
Mild convexity, 110
Minimal representation, 117
Minimax theorem, 159

SUBJECT INDEX

distance function, 210
inequality, 205
Mixed strategy, 157

Nearest-point property, 112
Neighborhood, 105
Network, 130
Node, 130
Norm, 2
Normal to a hyperplane, 31

Objective function, 168 1
Open ball, 4, 22
Open cover, 8
Open set, 4, 21
Optimal solution, 169, 220
Optimal strategy, 159
Optimization, 154
Origin, 2
Orthogonal, 2, 24

Parallel, 12, 31
Parallel body, 94
Parallelotope, 70, 128
Payoff, 157
Perimeter, 81
Point, 1, 19
  corner, 78
  equichordal, 92
  equilibrium, 155
  exposed, 45, 108, 116
  extreme, 42, 44, 45, 117, 121, 171
  interior, 4
  of mild convexity, 110
  saddle, 155, 220
  of strong local concavity, 108
  visible, 53
  of weak local concavity, 108
Polar set, 140, 142, 145, 211
Polygonally connected, 104
Polygonal path, 104
Polyhedral network, 130
Polyhedral set, 119, 120, 121, 170
Polytope, 19, 116, 119, 145
Positive combination, 23
Positive hull, 23, 46
Positively homogeneous function, 206
Primal programming problem, 173
Prism, 129, 143
Profile, 42, 43
Proper face, 116
Pure strategy, 157
Pyramid, 125, 146

Quasiconvex function, 205

Radon's theorem, 47, 54
Ray, 45
Real linear space, 1
Region of a network, 130
Regular polyhedron, 132
Regular polytope, 133
Relative interior, 12, 20
Relative topology, 4
Reuleaux triangle, 78, 83
Round, 113

Saddle point, 155, 220
Scalar multiple, 7
Schwarz inequality, 2
Section of a set, 107
Semi-norm, 214
Separated sets, 8
Separation:
  by cylinder, 64
  by hyperplane, 34, 36, 37, 39
  improper, 38
  by parallelotope, 70
  by slab, 57
  by spherical surface, 61
  strict, 34, 38, 39
  strong, 40
  weak, 34
Sequence, 6, 96, 98
Sequential compactness, 100
Simplex, 19, 20, 84, 124, 146
  method, 171, 183, 190
  tableau, 188
Simplicial property, 56, 59, 69
Skew-symmetric matrix, 168
Slab, 57, 64
Slack variable, 184
Smooth, 113
Solution to a game, 159
Star-shaped, 11, 25, 53, 101
Strategy, 157
Strictly convex, 84
Strictly feasible solution, 220
Strong local concavity, 108
Strong local support, 108
Strongly convex, 66
Strong universal cover, 85
Subadditive function, 212
Subset, 2
Subspace, 16
Support:
  function, 206

by half-space, 115
   by hyperplane, 41, 45, 77, 116
   strong local, 108
   weak local, 108
Surface area, 133, 136
Symmetric matrix game, 168

Tetrahedron, 19, 124
Theorem of alternative for matrices, 160
Theorem of supporting hyperplanes, 160
Tietze's theorem, 110
Topology, 4, 113
Translate, 7, 50, 55
Transversal, 55
Triangle, 19

Unbounded program, 170
Uniformly bounded, 97
Unit vector, 2
Universal cover, 85

Value:
   of game, 158, 159
   of strategy, 158
Vector, 1
Vertex, 19, 78, 116, 117
   figure, 133
Visible point, 53
Volume, 133, 136

Weak local concavity, 108
Weak local support, 108
Width, 76